Automated Hierarchical Synthesis of
Radio-Frequency Integrated Circuits and Systems

Fábio Passos • Elisenda Roca
Rafael Castro-López • Francisco V. Fernández

Automated Hierarchical Synthesis of Radio-Frequency Integrated Circuits and Systems

A Systematic and Multilevel Approach

 Springer

Fábio Passos
Instituto de Telecomunicações
Lisbon, Portugal

Rafael Castro-López
Instituto de Microelectrónica de Sevilla
CSIC and Universidad de Sevilla
Sevilla, Spain

Elisenda Roca
Instituto de Microelectrónica de Sevilla
CSIC and Universidad de Sevilla
Sevilla, Spain

Francisco V. Fernández
Instituto de Microelectrónica de Sevilla
CSIC and Universidad de Sevilla
Sevilla, Spain

ISBN 978-3-030-47249-8   ISBN 978-3-030-47247-4   (eBook)
https://doi.org/10.1007/978-3-030-47247-4

This Springer imprint is published by the registered company Springer Nature Switzerland AG
The registered company address is: Gewerbestrasse 11, 6330 Cham, Switzerland

To Inês, Carlos, André, and Rita

F. Passos

To Judit and Nuria

E. Roca and F. V. Fernández

To Alejandro and Martina

R. Castro-López

Preface

Radio-frequency (RF) circuits are of utmost importance in applications developed for the Internet of Things (IoT), the fifth-generation (5G) broadband technology and electronic health (eHealth) monitoring. The design of RF circuits in nanometric technologies for IoT/5G/eHealth applications is becoming extraordinarily difficult due to the high complexity and demanding performances of such circuits/systems. Therefore, traditional design methodologies based on iterative, mostly manual, processes are unable to meet such requirements. Moreover, current EDA tools are getting outdated because they only support that kind of traditional methodologies. Also, the short time-to-market demanded by nowadays IoT/5G/eHealth applications is creating a design gap, thus leading to a productivity decrease in the deployment of such applications. Therefore, new design methodologies such as the ones proposed in this book are in need in order to increase the designer efficiency in the deployment of such applications.

This book focuses on ways to develop new design methodologies that allow the optimization-based synthesis of RF systems in a seamless, efficient, and accurate way. The supported basic idea is to develop methodologies based on the bottom-up hierarchical multilevel approach, where the system is designed in a bottom-up fashion, starting from the device level. Furthermore, at each level of the design hierarchy, several aspects must be taken into account in order to increase the robustness of the designs and increase the accuracy and efficiency of the simulations. The methodologies proposed in this book are able to handle circuit sizing and layout in a complete and automated integrated manner, in order to achieve fully optimal designs in much shorter times than traditional approaches. Moreover, the methodology also takes into account process variability. Therefore, it is possible to say that the overall goal of the book is to propose an efficient and accurate methodology able to automatically design RF systems, in which the upmost accuracy has to be guaranteed from the device to the system level. Throughout the book, strategies are proposed in each level of the hierarchy (from device to system level), augmenting the accuracy, efficiency, and the ability to achieve optimal results. Hence, the efforts were focused on the following directions:

- Development of accurate models for passive devices;
- Accurate and efficient synthesis methodologies for passive devices;
- Accurate, optimal, and efficient synthesis methodologies for circuits;
- Development of bottom-up design methodologies between the device and the circuit level;
- Development of variability-aware bottom-up design methodologies for RF circuits;
- Development of layout-aware bottom-up design methodologies for RF circuits;
- Development of layout-variability-aware bottom-up methodologies for RF circuits;
- Development of multilevel bottom-up design methodologies from the device up to the system level.

In order to show the innovative contributions in practice, the methodology developed is explained throughout the book, where each chapter focuses on different aspects of RF design. The book is organized as follows:

Chapter 1 explains the problematic issues in traditional design methodologies, discusses the automatic circuit design state-of-the-art and the demands for an accurate and efficient RF automatic circuit design methodology.

Since the methodologies developed in this book are applied to RF blocks present in RF receivers, Chap. 2 reviews the traditional receiver architectures and the three main blocks that constitute the RF front-end (low noise amplifier (LNA), voltage controlled oscillator (VCO), and mixer) are also described, as well as its most important performance parameters.

Chapter 3 discusses the development of accurate models for passive devices, where a novel surrogate modeling strategy for integrated inductors is presented for an accurate and efficient modeling. The presented model shows less than 1% error when compared to EM simulations while reducing the simulation time by three orders of magnitude. Several models were created for different inductor topologies and the accuracy and efficiency of such models enable its usage within iterative optimization loops for inductor synthesis. The same chapter presents accurate and efficient synthesis methodologies for passive devices. Furthermore, the surrogate model developed was used with multi-objective optimization algorithms in order to achieve Pareto-optimal fronts (POFs), which provide the best possible trade-offs for the inductor performances, e.g., inductance vs. quality factor vs. area. Obtaining such inductor POFs in an efficient and accurate manner is a key ingredient for the successful development of the multilevel hierarchical methodology presented in the book. Furthermore, SIDe-O, an EDA tool for the design and optimization of integrated inductors is presented. This tool offers an intuitive graphical user interface (GUI), which can be used by any RF designer in order to model, simulate, and design integrated inductors for different topologies, operating frequencies, and technological processes. Besides, SIDe-O also allows the creation of S-parameter files that accurately describe the behavior of inductors for a given frequency range, which can later be used in electrical simulations for circuit design in commercial environments. The surrogate models developed, and integrated in SIDe-O, provide a solution to the problem of accurately and efficiently modeling inductors, as well as their optimization, alleviating the bottleneck that these devices represent in the RF circuit design process.

Chapter 4 regards the accurate, optimal, and efficient synthesis methodologies for RF circuits. In such chapter, a study is presented on how the inductor modeling error could impact the estimation of circuit performances. In order to do so, an inductor analytical model was compared against the previously developed surrogate model, and both were used in optimization-based circuit design approaches. It was proved that when used in the design/synthesis of RF circuits, analytical models, with typically high modeling errors, lead to suboptimal circuit designs, or, worse, to a disastrous non-fulfillment of specifications. Therefore, this states and reinforces the need for an accurate modeling of passive devices in RF, and therefore, the importance of the accurate inductor models illustrated in this book. Moreover, in all circuit optimizations, several simulation strategies were used in order to reduce the circuit simulation time and make optimizations more efficient. By using such strategies, some of the most expensive RF performances (e.g., third-order intercept point) can be efficiently calculated and considered during the automated design of RF circuits. In the same chapter, a wide study was carried out to compare two different optimization-based RF design methodologies: one based on hierarchical decomposition and bottom-up synthesis and another where synthesis was performed at the circuit level without hierarchical decomposition. In both cases, inductors were modeled using the same surrogate modeling strategy, and therefore, only efficiency and optimality were under comparison. It is demonstrated that bottom-up hierarchical design methodologies are far more efficient and are able to achieve superior results.

Chapter 5 presents a methodology where a layout-aware optimization is described which uses an automatic layout generation during the optimization for each sizing solution using a state-of-the-art module generator, template-based placer and router, which were specifically developed for RF circuits. The proposed approach exploits the capabilities of commercial layout parasitic extractors to determine the complete circuit layout parasitics. Also, in Chap. 5, the previously presented layout-aware methodology is further elaborated in order to take into account device process variability to develop a layout-corner-aware optimization.

Chapter 6 presents the multilevel bottom-up design methodologies which go from the device up to the system level. Such novel multilevel bottom-up circuit design methodology is described and applied to design an RF system composed of three blocks (an RF front-end composed of an LNA, a VCO, and mixer). This novel approach is disruptive because it covers the complete hierarchy of RF systems: starting at the device level, going up to the circuit level, and finally, reaching the system level. By using such multilevel bottom-up strategy, different circuits can be connected in order to build an RF system. Furthermore, each level of the hierarchy is simulated with the upmost accuracy possible: EM-level accuracy at device level and electrical simulations at higher levels. The methodology developed in this book encourages the hierarchical low-level POF reuse, which is typical in bottom-up methodologies. Moreover, the methodology proved to be highly efficient for the design of RF front-ends for different communication standards, and, compared to alternative synthesis strategies, the presented methodology shows superior results.

In the end, Chap. 7 draws some conclusions.

Lisbon, Portugal   Fábio Passos
Seville, Spain   Elisenda Roca
Seville, Spain   Rafael Castro-López
Seville, Spain   Francisco V. Fernández
February 2020

Contents

Chapter 1
Introduction

In an emerging telecommunications market, evolving towards 5G [1], it is estimated that there are over three billion smartphones users nowadays [2]. Only by itself, this number is astonishing. But nothing compares to what is going to happen in the foreseeable future. The next technological boom is directly related to the emerging internet-of-things (IoT) market. It is estimated that by 2020, there will be 20 billion physical devices connected and communicating with each other [3], which gives more than 2 physical devices per person on the planet. Due to this technological boom, new and interesting investment and research opportunities will emerge. In fact, it is estimated that in 2020 approximately three billion dollars will be invested in this market alone, 50% more than in 2017 [3]. Due to the fact that most of these IoT devices will have to communicate wirelessly among each other, and that radio-frequency (RF) circuits are essential for that purpose, there is, and there will be a high demand for RF circuits, nowadays and in the foreseeing years. Therefore, it is easy to understand why integrated circuit (IC) design companies specialized in RF, are already the companies which generate more income among all the fabless IC suppliers (e.g., Qualcomm and Broadcom, see Table 1.1).

The problem is that the design of RF circuits in nanometric technologies is becoming extremely difficult due to its increasing complexity. Designing an RF circuit is one of the most challenging tasks in nowadays electronics, partially due to its demanding specifications, convoluted trade-offs, and high operating frequencies. In fact, compared to its analog (baseband) and digital counterparts, the RF design requires a higher design effort despite the comparatively low number of devices (see Fig. 1.1). With todays' strict time-to-market restrictions and the need for design solutions with very demanding performance specifications, one of the areas where it is extremely important to focus is on the development of new systematic design methodologies for RF circuits. These RF circuit design methodologies must allow the designer to obtain circuits which comply with the demanding specifications in a reasonable time.

© Springer Nature Switzerland AG 2020
F. Passos et al., *Automated Hierarchical Synthesis of Radio-Frequency Integrated Circuits and Systems*, https://doi.org/10.1007/978-3-030-47247-4_1

Table 1.1 Illustration of the top 10 ranking of fabless IC suppliers for 2018 [4]

Rank	Company	Revenue
1	Broacom	18,941 M$
2	Qualcomm	16,370 M$
3	NVIDEA	11,163 M$
4	MediaTek	7882 M$
5	AMD	6475 M$
6	Xilinx	2868 M$
7	Marvell	2819 M$
8	Novatek	1813 M$
9	Realtek semiconductor	1518 M$
10	Dialog semiconductor	1443 M$

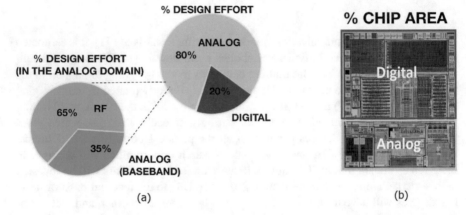

Fig. 1.1 (**a**) Illustration of the design effort comparison between analog and digital, and between analog (baseband) and RF. (**b**) Illustration of the area differences between the analog and digital parts

1.1 Traditional Design Methodologies

In the analog and RF domain, the traditional design methodology follows the flow illustrated in Fig. 1.2. The design of an IC starts by the definition of the circuit performances that have to be achieved, and then, the so-called electrical and physical synthesis have to be performed. These synthesis stages compose the core of any design flow and are the most important stages of any circuit design methodology. The first step of the flow is the electrical synthesis, where the designer must select an appropriate circuit topology and *size* the design. This *sizing* operation is a process where the designer finds the dimensions of each device used (transistors, capacitors, etc.) in order to meet the desired specifications. The output of this electrical synthesis is a schematic, which contains a list of all the devices composing the circuit and how they are connected. Furthermore, and more importantly, this schematic also includes the sizes of each single device (e.g., transistor lengths and

Fig. 1.2 Electrical and
physical synthesis

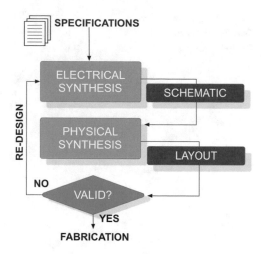

widths, etc.). After the electrical synthesis is performed, the physical synthesis must
be achieved. The goal of this step is to attain the physical representation of the
circuit, known as *layout*. This layout is a collection of geometric shapes and layers
which are later used for fabrication. After the physical synthesis, the layout of the
circuit must be verified, and, if valid, it is ready for fabrication. If not, some re-
design stages are needed.

If the circuit/system under design is too complex, analog/RF designers use
divide-and-conquer techniques in order to reduce the complexity of the entire
system. The basic idea is to partition the system into smaller pieces, which are easier
to manage. This is known as hierarchical partitioning. The most well-known hierarchical design strategies are the top-down and bottom-up design methodologies, as
shown in Fig. 1.3.

In top-down design methodologies, the designer starts by designing the system
level, and the performances are consecutively derived for the lower levels, until
reaching the device level. The circuit is designed in a more "abstract" way in high-
levels, relying in e.g. behavioral simulations, and, at lower levels, more precise
simulations can be performed. Furthermore, at each level of the design hierarchy,
a verification must be performed in order to check if the design is valid. One of
the advantages of top-down methodologies is that the performances for the entire
system are known since an initial design stage (although only estimated). However,
if any of the circuits composing the system do not attain the necessary performances,
some re-design iterations are needed in order to achieve the desired specifications.
In the worst possible scenario, the complete system architecture must be changed.

On the other hand, in bottom-up design methodologies, the design stage starts in
the device level and ends up in the system level. The main disadvantage of bottom-
up methodologies is that the system performances are only verified when all its
composing blocks are designed, which can lead to major design changes later in the
design process.

Fig. 1.3 Top-down vs.
bottom-up design
methodologies

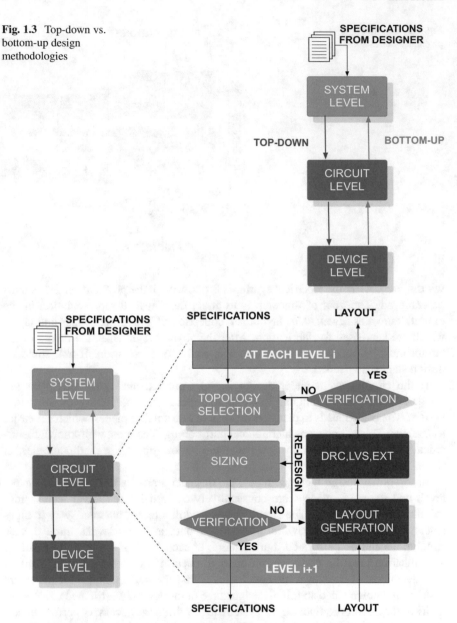

Fig. 1.4 General design flow for analog and RF integrated circuits

In practice, in traditional design methodologies, at each level of the hierarchy, the designer must perform a top-down electrical synthesis and a bottom-up physical synthesis, both needing a verification stage, as shown in Fig. 1.4.

As part of the electrical synthesis, the designer must select the architecture/topology which is capable of achieving the desired specifications. Afterwards, the sizing process is performed. At higher levels, the sizing is the process of mapping the current level specifications into the needed specifications for the immediately lower level. At device level, sizing is the process of dimensioning each passive and active circuit component. After the topology selection and sizing operation, the design is simulated and verified in order to check if the specifications are met. If the specifications are met, the flow continues to the next level.

The physical synthesis involves the layout generation stage, where the layout of a device, or circuit is generated. Afterwards, the layout is checked against a set of technology-defined rules with a design-rule check (DRC), and a layout-versus-schematic (LVS) check is performed, and if both checks are valid, the layout is acceptable for fabrication. Then, a parasitic extraction (EXT) must be performed. This is important in order to extract the layout-induced effects. These layout-induced effects add a set of *parasitic* capacitances, resistances, and inductances to the circuit, and therefore, may change its performances. If the specifications are not met after the layout extraction, the layout must be improved or, in a worse scenario, a re-sizing operation must be performed. The illustration of the complete hierarchical design for the levels of abstraction previously discussed (system, circuit, and device), is shown in Fig. 1.4.

All steps of the hierarchical flow shown in Fig. 1.4 can undertake several re-design iterations in order to reach the final system design that meets all specifications, therefore making the process of designing an IC a long and (usually) repetitive task. Hence, in order to relieve the designer from these long and repetitive tasks, the IC design process can be automated. In an ideal scenario, designers would have an electronic design automation (EDA) tool that could automatically perform the steps demonstrated in Fig. 1.4, something defined as a silicon compiler [5]. With this ideal tool, the user would only stipulate the desired specifications for his/her system and the tool would automatically generate the IC ready for fabrication. However, such a tool does not exist. In the digital domain the automatic circuit design tools are relatively close to the previously described silicon compiler. However, in the analog domain, and especially in RF, this silicon compiler is yet nothing but a dream.

Therefore, this book presents new systematic design methodologies capable of improving the state-of-the-art and cut short the distance between the RF and digital automatic design tools. By doing so, it will be possible to shorten the existing design productivity gap in RF circuit design.

In Sect. 1.2 a brief historical background on automatic circuit design is performed, and the current state-of-the-art is overviewed. In order to establish a new design methodology for RF systems, different bottlenecks of the RF design process must be addressed in order to successfully design such circuits. Hence, in Sect. 1.3, the demands for an accurate RF system design methodology are discussed.

1.2   Automatic Circuit Design State-of-the-Art

In this section, the state-of-the-art on automatic circuit design methodologies is reviewed. As previously mentioned, the electrical and physical synthesis are the core of any design methodology and, therefore, the state-of-the-art for both of them is reviewed.

1.2.1   Knowledge-Based Approaches

The basic idea of knowledge-based approaches is to have a pre-defined design plan, in the form of design equations or design strategies, to find the circuit sizing/layout so that the specifications are met. These type of tools are known as knowledge-based approaches because they use knowledge and expertise from the designer in order to establish/define a design plan for a given circuit.

1.2.1.1   Knowledge-Based Electrical Synthesis

In the 1990s, several tools were developed which could automatically perform electrical synthesis of analog circuits [6–10]. In these tools, the design plan was basically a set of analytical equations, which were used to solve the circuit. The tool provided the means to automatically execute a routine that would solve all the equations and, therefore, size the circuit under study. The main advantage of these approaches is its short execution time. However, deriving the design plan is hard and time-consuming, the derived equations are usually too simple and do not incorporate all the device physics. Moreover, the design plan requires constant maintenance in order to keep it up to date with technological evolution, and the results are not optimal, and only suitable for a first-cut design.

1.2.1.2   Knowledge-Based Physical Synthesis

In order to perform circuit physical synthesis, other knowledge-based tools were also developed. Roughly, the layout generation phases are *placement*, where all circuit components are distributed over the layout plane (also called floorplan), and routing, where all components are interconnected. Automatic knowledge-based layout generation tools were developed in order to generate the circuit layout in such a way that placement and routing were specified in advance. There are two types of knowledge-based approaches for automatic layout generation: rule-based and template-based approaches. Rule-based approaches use a set of rules that have to be followed by whichever placement and routing algorithms used during circuit layout generation [11]. In template-based approaches, the main idea is to capture the designer expertise in a template that specifies all necessary component floorplanning and the routing spatial relationships. Moreover, the template must capture analog specific constraints like routing symmetry and device matching [12].

1.2.2   Optimization-Based Approaches

Knowledge-based design tools were developed in order to automatize some of the tasks inherent to analog/RF designers, without aiming at optimality. In order to reach optimal designs, optimization algorithms can be used in order to perform electrical/physical synthesis. The design of any circuit/system can be posed as an optimization problem, mathematically defined as,

$$\text{minimize} \quad f(x); \quad \text{f}(x) = \{f_1(x), f_2(x), \ldots, f_n(x)\} \in \mathbb{R}^n$$

$$\text{such that} \quad g(x) \geq 0; \quad \text{g}(x) = \{g_1(x), g_2(x), \ldots, g_m(x)\} \in \mathbb{R}^m \qquad (1.1)$$

$$\text{where} \quad x_{Li} \geq x_i \geq x_{Ui}, i \in [1, p]$$

where x is a vector with p design parameters, each design parameter being restricted between a lower limit x_{Li} and an upper limit x_{Ui}. The functions f_j, with $1 \leq j \leq n$, are the objectives that will be optimized, where n is the total number of objectives. The functions g_k, with $1 \leq k \leq m$, are design constraints. The basic approach to solve Eq. (1.1) is illustrated in Fig. 1.5. It is possible to observe that the optimization algorithm is linked with a performance estimator, where the designer chooses the circuit performances to be considered (optimization objectives and constraints) and executes the algorithm which then returns the circuit sizing (e.g., widths and lengths of transistors).

Fig. 1.5 Optimization-based methodology for circuit design

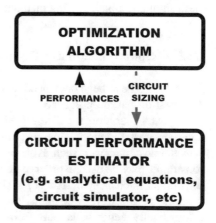

1.2.2.1 Optimization-Based Electrical Synthesis

While performing optimization-based electrical synthesis, there are two main categories, namely, equation-based and simulation-based.

The equation-based methods use analytical equations in order to evaluate the circuit performances. Several tools were developed which implemented this method [13–21]. Equation-based optimization-based sizing is similar to the knowledge-based sizing methods in the sense that they both use relatively simple analytical equations in order to estimate the circuit performances. However, equation-based methodologies do not need an explicit "design plan" to be defined. Also, the methods presented in this section go a step further by linking the equations with optimization algorithms, which were developed in order to reach optimal results. Similarly to the knowledge-based approaches, the advantage of equation-based methods is the short evaluation time. These methods are extremely suitable to find first-cut designs. However, like the knowledge-based approaches, the main drawback is that not all physical characteristics of the devices can be easily captured by analytic equations, making the method inaccurate (especially for RF circuits) and the generalization to different circuits, technologies, and specifications is very time-consuming because new equations must be derived.

Equation-based optimization methodologies are suitable because they are computationally cheap and, therefore, very fast to evaluate. However, they lack sufficient accuracy. Therefore, instead of using analytical equations in order to estimate the circuit performances, a circuit simulator (e.g., electrical simulator [22]) should be used in order to accurately estimate the circuit performances. The advantage is that these type of simulators tend to be much more accurate than analytical equations. The methods linking an optimization algorithm with a circuit simulator are usually defined as simulation-based strategies. Therefore, in order to obtain more accurate designs, this simulation-based optimization gained ground and became the most common optimization-based strategy. Some of the developed works that employ these simulation-based sizing methods can be found in [23–37].

1.2.2.2 Optimization-Based Physical Synthesis

Several tools have been developed that are able to perform physical synthesis using optimization-based approaches. With such tools, placement and routing stages of the layout generation are determined by an optimization algorithm according to a certain cost function. This cost function typically considers the minimization of some design aspect, such as, layout area or routing length. Furthermore, some constraints may be used in order to penalize the violation of some analog/RF design constraints, such as symmetrical RF signal paths, device mismatch, etc.

Some of the developed tools, the so-called heuristic approaches, are able to automatically generate layouts from circuit descriptions, while handling typical analog layout constraints such as, device matching, symmetry, etc. However, these approaches do not account for the performance degradation that appears due to the physical implementation of the devices [38]. Therefore, they do not provide promising results because the layout parasitic effects, which highly degrade the performances of the circuits, are not taken into account during the design stage. Therefore, one of the keys in order to have a successful circuit synthesis is that the electrical synthesis and the physical synthesis should not be considered as separate steps of the design methodology. Hence, new optimization-based physical synthesis approaches appeared, the so-called performance-driven. In these approaches, the layout-induced effects are taken into account [39]. These performance-driven tools try to measure the layout-induced degradation and keep it below desired margins. Thus, the impact of each layout parasitic is weighed out according to its effect on the circuit performance.

1.2.3 Hierarchical Optimization-Based Approaches

Similarly to what happens in the traditional circuit design flow, the divide-and-conquer techniques can also be used in optimization-based methodologies in order to ease the optimization and, therefore, the design process. This kind of divide-and-conquer techniques are particular useful, because, when the problem is too complex (e.g., too many design variables), optimization algorithms struggle to converge to optimal solutions, and the process can become inefficient. Therefore, the previously described top-down and bottom-up design strategies can be applied to optimization-based design methodologies.

The top-down design methodologies are illustrated in Fig. 1.6. In top-down design methodologies, the designer sets the specification for the highest level (e.g., system level). During the high-level optimization, the "design variables" are the performances for the lower-level circuits. After obtaining the high-level design, the

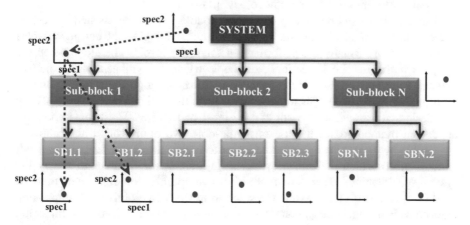

Fig. 1.6 Top-down design methodologies

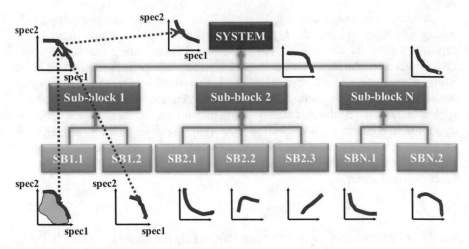

Fig. 1.7 Bottom-up design methodologies

performances for the lower levels (e.g., Sub-block 1, Sub-block 2, etc.), must be attained, and this process continues down to the lowest level of the hierarchy (e.g., SB1.1, SB1.2, etc.). Several works used this type of design methodologies [31, 33].

The problem with this kind of methodologies is that the designer commonly uses a system-level design tool in order to obtain the specifications of lower-level blocks. Afterwards, when the designer tries to synthesize the lower-level blocks, it may happen that some of the needed specifications are impossible to meet and therefore, re-design cycles are needed, which will degrade the efficiency of the entire process.

In order to reduce, or even eliminate, the re-design cycles, bottom-up design methodologies can be used. Figure 1.7 illustrate this type of methodologies. The main idea in bottom-up methodologies is to start designing the system from the circuit level (e.g., SB1.1, SB1.2, etc.) until reaching the system level. Several different works employed this methodology [34–37].

Both top-down and bottom-up design methodologies can be assisted by several different optimization algorithms. In top-down methodologies, the design variables at each level are usually the performances for lower levels. Therefore, when the designer optimizes the lower level, he/she is trying to synthesize those performances. In order to do so, single-objective optimization algorithms are commonly used ($n = 1$ in Eq. (1.1)). Therefore, at each level the designer would achieve only one design (illustrated by the dot at each level of Fig. 1.6). However, in bottom-up design methodologies the use of single-objective optimization algorithms is impossible because, when going from lower to higher levels the designer does not know a priori which performances he/she is looking for in order to satisfy the system specifications. Therefore, multi-objective optimization algorithms must be used ($n > 1$ in Eq. (1.1)). While the solution to the single-objective optimization algorithms is a single design point, the solution to the multi-objective optimization algorithms is a set of solutions exhibiting the best trade-offs between the objectives (illustrated by the curve at each level of Fig. 1.7, for a case with two objectives).

Therefore, when synthesizing a given high-level block, the design space of that optimization is set by the designs available from lower levels. In practical terms, what the designer is doing with these methodologies, is exploring the design space of each level and finding the optimal designs for that level, hence, building an optimized library for each device/circuit/system.

When discussing multi-objective optimization algorithms, a few key concepts must be first established. In single-objective optimization algorithms, the final obtained solution can be considered the best one because it is the one that achieved the "best" value for the objective function value $f(x)$. However, for multi-objective optimization this cannot be performed because there are several objectives. Therefore a new concept must be established. This concept is denoted as Pareto dominance. A design point a is considered to *dominate* the design point b, if $f_j(a) \leq f_j(b) \ \forall j$ and $f_i(a) < f_i(b)$ for at least one objective i (for minimization problems). The design point a is said to be non-dominated if there is no other design point that dominates it. The non-dominated set of the entire feasible[1] search space is known as Pareto set, exhibiting the best trade-offs between the objectives, i.e., the Pareto-optimal front (POF). The concept of Pareto dominance and the Pareto-optimal front are described in Fig. 1.8 for a problem where both $f_1(x)$ and $f_2(x)$ are minimized. It is possible to see that y^a is non-dominated and Pareto optimal because $f_1(y^a)$ and $f_2(y^a)$ are lower than $f_1(y^b)$, $f_1(y^c)$ and also $f_2(y^b)$, $f_2(y^c)$.

These multi-objective optimization algorithms are extremely useful since the circuit sizing is in its essence a multi-objective optimization problem, and the

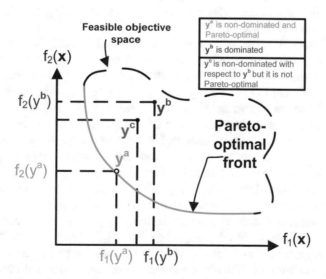

Fig. 1.8 Illustrating Pareto dominance and Pareto-optimal front concepts for a 2-dimensional performance space

---

[1] A feasible point is a point that complies with the constraints of the optimization problem.

designer often wants to explore the trade-offs among conflicting performances, for example, the power consumption versus gain of a low noise amplifier.

1.3   Demands for an Accurate RF Automatic Circuit Design Methodology

From all the cited works so far, most of them are centered on the analog and mixed-signal domain, and only a few of them found its path to the RF world (e.g., [28, 32, 34, 36]). The design (and consequently automation) of RF circuits is far more complex and delicate than the analog ones due to several reasons. The difference between baseband and RF is largely due to the fact that capacitive and inductive reactances tend to be more significant at high frequencies than they are at lower frequencies. At the lower frequencies, those reactances exist, but they can usually be ignored. On the other hand, at RF frequencies, the parasitic or distributed reactances tend to be significantly higher. Another fact that highly affects the automation of RF circuit design is the time and accuracy of the analysis needed. The analyses needed to estimate the RF circuit performances are more time-consuming than the ones used in the analog baseband domain (e.g., AC analysis vs. periodic steady-state analysis). Furthermore, techniques such as periodic steady-state (PSS) have only been available in the last few years, leaving early RF designers with limited options such as transient analysis, which is highly time-consuming. Also, in the past, the available techniques used in order to calculate performances e.g. circuit noise, were fuzzy and sometimes inaccurate [32].

Even in modern times, with modern commercial SPICE-like simulators, the efficiency of optimization-based approaches can be hampered due to convergence issues in analysis such as PSS, and, performances such as input-referred third-order intercept point (IIP_3) usually need power sweeps, which highly decrease the efficiency of the optimization process. Therefore, it can be concluded that developing an efficient automated circuit design methodology is not an easy task. In order to develop an accurate automated RF circuit design methodology some of the biggest bottlenecks are: the accurate modeling of passive components such as integrated inductors, the layout parasitics that are very important in RF frequencies, and the process variability which highly affects nanometer technologies. Another important issue is the circuit complexity that the methodology is able to cope with. Most of the automated circuit design methodologies presented in literature are only suited for the RF block level design. When the system level is reached, designers usually use high-level simulation tools in order to estimate performances. However, these high-level tools may introduce some deviation from an actual transistor-level simulation. This fact is even more important in RF, because accuracy is a must.

In the next sections the issues that must be tackled in order to develop an accurate and efficient automated RF IC design methodology are discussed in detail.

1.3.1 Circuit Performance Evaluation

In RF design, evaluating the circuit performances with analytical equations is not a valid approach because they are simply not sufficiently accurate. RF circuits are extremely sensitive to any performance deviation, which may cause the circuit to malfunction. Therefore, RF circuit simulators must be used, such as SpectreRF [40], EldoRF [41], or HspiceRF [42]. Commercial SPICE-like circuit simulators are probably the most established CAD tool in the RF design flow, being used to verify the performance of the circuit since the early design stages until post-layout validations. Therefore, this is a mandatory requirement for high accuracy. However, some analysis are lengthy and can have convergence problems, as mentioned before. Therefore, the designer must use/develop efficient simulation strategies for each performance.

1.3.2 Integrated Inductor Modeling and Synthesis

In nowadays RF ICs, passive components play a key role in circuit design for impedance matching, tuning, filtering, or biasing. For example, it is estimated that in a cellular phone, passive components account for 90% of the component count, 80% of the size, and 70% of the cost [43]. From all passive components (e.g., resistors, capacitors, and inductors), while resistors and capacitors are accurately modeled in CMOS technologies, inductors are still a bottleneck for designers. Several authors discussed the inclusion of inductors' performances during their optimization methodologies using several different strategies. The most straightforward option is to use foundry-provided inductor libraries/models, as performed in [28, 36, 44] (see Fig. 1.9a).

However, these models usually do not provide sufficient accuracy for these passive components. Furthermore, if an inductor library is provided by the foundry, it is usually a limitative option because it reduces the possibility of finding an optimal inductor for a given application. Therefore, some authors fancy using simulators/models that are able to relate performance parameters with the inductor geometric parameters, which provide a wider range of inductor choices. The most accurate inductor evaluator (electromagnetic (EM) simulator) was used in [45] (see Fig. 1.9b). However, EM simulations are very time-consuming and, therefore, including them in an optimization-based process, where thousands of simulations must be performed, makes it an inefficient and unfeasible option. Moreover, the circuit designed in [45] has only one inductor. If a circuit with more inductors is needed, the number of EM simulations would increase, converting this method into an unaffordable one. On the other hand, analytical/physical models are able to calculate inductor performances very efficiently. In [46] a compact model is used to incorporate the inductor performances during the optimization of RF power amplifiers. Similar methodologies use inductor analytical models, such as the π-

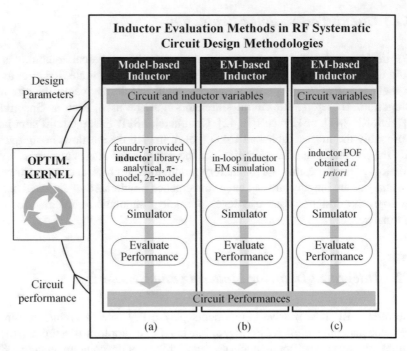

Fig. 1.9 Differences between inductor evaluation techniques in systematic circuit design method-ologies ((**a**) Model-based inductor, (**b**) EM-based inductor) and bottom-up systematic circuit design methodologies ((**c**) EM-based inductor)

model presented in [47] and illustrated in Fig. 1.10 [30, 48, 49] or the 2π-model [50] (see Fig. 1.9a). However, most analytical models do not present sufficient accuracy, especially at high frequencies [51]. As a way of achieving EM accuracy but avoiding EM simulations during the optimization of a given circuit, in [34, 52] a Pareto-optimal front (POF) of EM simulated inductors is obtained prior to any circuit optimization, and then, the inductor POF is used as optimal design space during a given circuit optimization (see Fig. 1.9c). By doing so, the inductors are modeled with EM accuracy and no EM simulation is performed during a circuit optimization, reducing therefore the total circuit design time. Furthermore, the POF has to be generated only once for a given inductor topology and operating frequency, and can later be used in several circuit optimizations. However, even though the inductor POF generation is only performed once for a given topology and operating frequency, the generation of the POF could still take several days or weeks. Hence, if a new inductor topology is needed, or the circuit operating frequency changes, a new inductor POF has to be generated, which is a very lengthy process.

In the last few years, surrogate inductor models have arisen as an attractive alternative aimed at combining the efficiency of analytical models with the accuracy of EM simulation [51]. Surrogate models can be global or local. The former ones try to construct a high-fidelity model that is as accurate as possible over the complete

Fig. 1.10 Typical integrated inductor physical model (π-model)

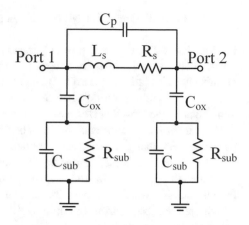

search space. Once the model is built, it can be used as a fast performance evaluator in an optimization algorithm for inductor synthesis [53]. However, it has been reported that these models may be highly inaccurate in some regions of the design space, yielding suboptimal results [54]. On the contrary, local models are iteratively improved during the inductor optimization process [55]. An initial coarse model is first created by using a few training points electromagnetically simulated. Then, this coarse model is used within a population-based optimization algorithm and, at each iteration, promising solutions (typically one) are simulated electromagnetically. The data from these EM simulations are used to improve the accuracy of the surrogate model in the region where the new simulation points are added, while evolving towards the presumed optimal inductor. However, the results may highly depend on the accuracy of the initial coarse surrogate model. A prescreening technique, e.g., the expected improvement (EI) method, which can be used to increase the quality of the optimization process, consists in using the uncertainty measurement of the prediction, i.e., the mean square error (MSE), instead of just the predicted value to rank promising solutions. These methods have been widely applied to single-objective optimization [56] and other works have tried to extend these approaches to the multi-objective case [57–60].

1.3.3   Taking into Account Layout Parasitics

Nowadays, the circuit sizing automation by means of optimization-based techniques is an established concept. However, in order to achieve robust circuit designs, complete circuit layout parasitic effects have to be considered during the automatic flow. This is even more critical for RF ICs where the impact of the layout parasitics is highly destructive due to the high operating frequencies.

1.3.3.1   Parasitic-Inclusive Methodologies

During the past few years, several parasitic-inclusive methodologies were developed. These methodologies tried to shorten the gap between schematic and physical circuit implementations. Parasitic-inclusive methodologies are approaches that use performance/symbolic models in order to estimate the impact of the layout parasitics and calculate the circuit performances. Some works, such as [61, 62], use symbolic models in order to estimate the effects of critical interconnections and layout parasitics. By using performance models, the layout parasitic estimation may sometimes be inaccurate. These approaches, based on performance models, are illustrated in Fig. 1.11a. In [63–65] the parasitics are extracted from a first coarse layout, and afterwards this parasitic information is used in order to create models which are then used during the optimization to estimate the performance of given solutions. The problem is that the parasitic information associated with a single layout design (or a reduced set) does not capture all parasitic variations that could be found during sizing, and, therefore, promising solutions may be lost. These approaches, based on parasitic sampling, are illustrated in Fig. 1.11b.

None of the methodologies presented in Fig. 1.11a and b performs an explicit layout generation during the optimization flow for each tentative sizing. The methodologies that are able to create a layout for each sizing solution during the

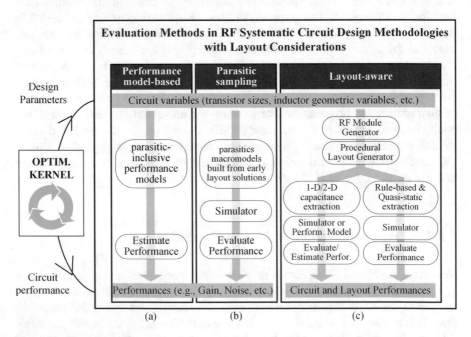

Fig. 1.11 Illustrating different strategies to include parasitics in optimization-based methodologies

optimization loop are designated as layout-aware methodologies, and are described in the following sub-section

1.3.3.2   Layout-Aware Methodologies

While layout generation in-the-loop (Fig. 1.11c) represents an overhead during optimization, having the layout readily available allows the computation of the precise parasitics for each specific solution without approximations.

In [66, 67] RF-specific methodologies were developed. In these works, a tool that is able to generate the physical layout of a device by means of an automatic routine is used. Such tool is defined as a module generator (MG) [68]. In [66], this MG was integrated into an automatic layout generator that uses automatic routines in order to place the devices in a previously described and always fixed position.[2] The tools that follow the same procedure/routine to perform the circuit layout are defined as procedural layout generators. Procedural generators are used to create the layout of each individual solution during the optimization, and, while in [66] only the parasitics of the critical nets were extracted using 1D/2D capacitance models, in [67], a more extensive set of parasitics are obtained using standard rule-based and quasi-static inductance extraction techniques. These techniques are illustrated in Fig. 1.11c.

It can be concluded that the only RF-specific methodologies that perform an automatic layout during the optimization loop adopt limitative procedural layout generators. Therefore, in order to obtain a more dynamic approach, instead of the procedural layout generators, template-based approaches should be considered [69–72]. In these template-based approaches the designer defines a layout template (or multiple templates) for a given circuit which may include a complete description of the floorplan.

For parasitic extraction, the analytical/geometrical 1D/2D, rule-based, and quasi-static methods present satisfactory results for baseband; however, the accuracy of such extractions is inferior when compared to the accuracy of 2.5D rule-based techniques or 3D field solvers. Therefore, for the upmost accuracy, these 2.5D rule-based techniques or a 3D field solver should be used.

1.3.4   Taking into Account Process Variability

When designing a circuit, the designer must take into account that some variation will occur between the simulated and the fabricated design. Variation is a huge problem in nanometer technologies, and failing to effectively take into account these variations can cause re-design iterations which ultimately result in product delays. These are serious issues that directly impact the revenues, profits, and ultimately,

---

[2]E.g., device A placed left of device B, device C on top of device A and B.

valuations of semiconductor companies and foundries alike. The variation causes may take many forms: environmental variations, such as temperature, power supply voltage, etc., or process and mismatch variations. While the environmental variations affect the circuit after its fabrication, process and mismatch variations are introduced during manufacturing, by random dopant fluctuations and other manufacturing problems (e.g., lithography). The process variations are inter-die, meaning that they affect all dies equally. On the other hand, the mismatch is an intra-die phenomenon, which means it affects devices in the same die. Several automatic design methodologies have been proposed that incorporate the variability effects into their flow in different ways. Some approaches estimate the circuit performances at its performance "corners." A corner is a point in the performance variation space, which represents the (supposedly) bounds of the model parameters. These corners enable a fast strategy to include process variability in automatic design methodologies. However, by considering only the device corner performances, the designer does not have an insight on the mismatch and on an important measure in variability-aware methodologies: the yield. The yield is the percentage of manufactured circuits that meet specs across all environmental conditions, expressed as a percentage e.g., 95% [73]. In order to have a yield estimation, the designer must perform a circuit statistical analysis, such as Monte-Carlo. However, this analysis involves hundreds of simulations which is a very time-consuming process. This process gets even worse when optimization-based methodologies are used, where thousands of Monte-Carlo simulations would have to be performed. In [74], a variability-aware methodology is considered by calculating the circuit performances at the nominal and performance corners of each tentative sizing solution. By doing so, the designer guarantees that his/her design will work even in the most pessimistic situations. In [75], a tool for the automated variation-aware sizing of analog integrated circuits is presented. This tool allows nominal, environmental, and process corner simulation in order to estimate the variability of the circuit. Furthermore, the tool uses response surface models in order to speed up the optimization. In [76], a method to calculate the trade-off between the yield and a figure of merit is presented. A quasi-Monte-Carlo sampling is performed in order to calculate the yield in a more efficient way. In [77], a technique based on artificial intelligence is used in order to speed up the yield optimization. However, this solution is implemented using single-objective algorithms rather than multi-objective. In [78], an efficient yield optimization technique for multi-objective optimization-based automatic analog integrated circuit sizing is presented. The proposed yield estimation technique reduces the number of required Monte-Carlo simulations by using the k-means clustering algorithm, with a variable number of clusters, to select only a handful of potential solutions where the Monte-Carlo simulations are performed.

There is an efficiency/over-design trade-off between the available methodologies to consider the process variations. While considering performance corners is an efficient solution, performing Monte-Carlo simulations is a more time-consuming option. The trade-off appears because if the designer considers the performance corners, he/she may be over-designing the circuit. It should also be taken into account that by using corners the mismatch between devices is also not considered.

1.3.5   Circuit Complexity

In the previous sections, the needs for an automated and efficient RF circuit design methodology were presented. However, it is also important to discuss the circuit complexity that the methodology is able to cope with.

In the past, most efforts directed to system level[3] design were focused in the direction of high-level system specification tools, RF budget analyzers, and architecture comparison tools [33, 79–81]. These high-level tools are very handy in top-down design methodologies because they are used in order to estimate lower-level circuit performances in order to fulfill the complete system specifications. After specifying the needed performances for each circuit, and, following a top-down approach, the designer starts from the higher level until reaching the lowest possible level (e.g., synthesizing the passive components). However, while going down the hierarchy, the designer may realize it may be difficult or even impossible to synthesize the needed performances for some circuit/passive, which ultimately will lead to unwanted re-design cycles. Also, these high-level tools do not consider all circuit nonlinearities and, therefore, it may be difficult to guarantee that the specifications given by the tool will be met at the transistor level, which again, can lead to re-design cycles. Eventually, since the high-level specifications do not entirely match the transistor specifications, the designer can choose to over-design the RF system in order to reduce the re-design cycles. However, this would ultimately lead to suboptimal designs (e.g., circuits with higher power consumption). In [82], S. Rodriguez et al. created a tool that was able to automatically size each circuit of an RF system based on system specifications. In this work, the entire system is optimized at once with no top-down or bottom-up hierarchy. By using an optimization engine, each component of each circuit composing the system would be automatically sized. However, the tool presents some drawbacks: the circuit performances are estimated using analytical equations during the optimization stage and ideal models are used for the passive components. By using analytical equations, the optimization is very efficient; however, the circuit performances may change significantly from an actual electrical simulation. Hence, it is reported in the paper that a fine-tuning operation must be performed in order to meet the desired specifications after the optimization. One of these fine-tuning operations is portrayed, where some components had to be changed by more than 50% from its initial value. In [83], Z. Pan et al. build performance models for each circuit of an RF front-end receiver and a Verilog description of these circuits is constructed in order to simulate the circuit performances. While for complex systems, such as, e.g., an RF transceiver, full system simulation is not practical to perform using an electrical simulator, for smaller systems such as a receiver, this simulation is still manageable. Therefore, electrical simulators should be preferred instead of performance models, in order to achieve superior

---

[3]In this book we denote by *system* any set of two different circuits that are connected.

accuracies. Thus, in [83], the system design is efficient but when the design is actually simulated at transistor level, the performances are expected to change due to the usage of approximated performance models. Optimizing the system entirely as in [82] is an inefficient solution because RF systems may be highly complex with a huge set of design variables and specifications, which can degrade the convergence of the optimization algorithm. On the other hand, using analytical equations and performance models as in [83] will inevitably lead to inaccurate designs. Therefore, it can be concluded that automating the design of an RF system and accurately calculating its performances is not a straightforward task. Such scenery gets even more obvious, because from the several works that use transistor-level simulations together with optimization-based methodologies for the automated design of RF circuits (e.g., [29, 30, 34, 36, 45, 46, 62, 84–86]) only M. Chu in [86] connected an LNA and a mixer. This fact is due to the high number of design variables, circuit/system specifications, and the difficulties to efficiently calculate the system performances.

1.4 Summary

In this Chapter the problematic issues in traditional design methodologies were presented and the main differences between top-down and bottom-up design methodologies are explained. Furthermore, the state-of-the-art for the automatic circuit design has been discussed by covering different approaches that were used throughout the years. Furthermore, a number of key demands are presented in order to achieve an accurate and efficient RF automatic circuit design methodology, able to design RF systems.

References

1. P. Rost, A. Banchs, I. Berberana, M. Breitbach, M. Doll, H. Droste, C. Mannweiler, M.A. Puente, K. Samdanis, B. Sayadi, Mobile network architecture evolution toward 5G. IEEE Commun. Mag. **54**, 84–91 (2016)
2. https://www.statista.com/statistics/330695/number-of-smartphone-users-worldwide. Accessed 28 Jan 2020
3. http://www.gartner.com/newsroom/id/3598917. Accessed 7 Feb 2018
4. https://press.trendforce.com/press/20190227-3216.html. Accessed 29 Jan 2020
5. S.L. Hurst, *VLSI Custom Microelectronics: Digital, Analog, and Mixed-Signal* (Dekker, New York, 1999)
6. M.G.R. Degrauwe, O. Nys, E. Dijkstra, J. Rijmenants, S. Bitz, B.L.A.G. Goffart, E.A. Vittoz, S. Cserveny, C. Meixenberger, G. van der Stappen, H.J. Oguey, IDAC: an interactive design tool for analog CMOS circuits. IEEE J. Solid State Circ. **22**, 1106–1116 (1987)
7. R. Harjani, R.A. Rutenbar, L.R. Carley, OASYS: a framework for analog circuit synthesis. IEEE Trans. Comput. Aided Des. Integr. Circuits Syst. **8**, 1247–1266 (1989)

8. N.C. Horta, J.E. Franca, Algorithm-driven synthesis of data conversion architectures. IEEE Trans. Comput. Aided Des. Integr. Circuits Syst. **16**, 1116–1135 (1997)
9. F. El-Turky, E.E. Perry, BLADES: an artificial intelligence approach to analog circuit design. IEEE Trans. Comput. Aided Des. Integr. Circuits Syst. **8**, 680–692 (1989)
10. C.A. Makris, C. Toumazou, Analog ic design automation. II. automated circuit correction by qualitative reasoning. IEEE Trans. Comput. Aided Des. Integr. Circuits Syst. **14**, 239–254 (1995)
11. V.M. zu Bexten, C. Moraga, R. Klinke, W. Brockherde, K.G. Hess, ALSYN: flexible rule-based layout synthesis for analog IC's. IEEE J. Solid State Circ. **28**, 261–268 (1993)
12. B.R. Owen, R. Duncan, S. Jantzi, C. Ouslis, S. Rezania, K. Martin, BALLISTIC: an analog layout language, in *Proceedings of the IEEE Custom Integrated Circuits Conference* (1995), pp. 41–44
13. H.Y. Koh, C.H. Sequin, P.R. Gray, OPASYN: a compiler for CMOS operational amplifiers. IEEE Trans. Comput. Aided Des. Integr. Circuits Syst. **9**, 113–125 (1990)
14. G.G.E. Gielen, H.C.C. Walscharts, W.M.C. Sansen, Analog circuit design optimization based on symbolic simulation and simulated annealing. IEEE J. Solid State Circ. **25**, 707–713 (1990)
15. G.G.E. Gielen, H.C.C. Walscharts, W.M.C. Sansen, ISAAC: a symbolic simulator for analog integrated circuits. IEEE J. Solid State Circ. **24**, 1587–1597 (1989)
16. K. Swings, W. Sansen, DONALD: A workbench for interactive design space exploration and sizing of analog circuits, in *European Conference on Design Automation* (1991), pp. 475–479
17. P.C. Maulik, L.R. Carley, D.J. Allstot, Sizing of cell-level analog circuits using constrained optimization techniques. IEEE J. Solid State Circ. **28**, 233–241 (1993)
18. E.S. Ochotta, R.A. Rutenbar, L.R. Carley, Synthesis of high-performance analog circuits in ASTRX/OBLX. IEEE Trans. Comput. Aided Des. Integr. Circuits Syst. **15**, 273–294 (1996)
19. F. Medeiro, B. Pérez-Verdu, A. Rodriguez-Vázquez, J.L. Huertas, A vertically-integrated tool for automated design of sigma-delta modulators, in *European Solid-State Circuits Conference* (1994), pp. 164–167
20. A. Doboli, N. Dhanwada, A. Nunez-Aldana, R. Vemuri, A two-layer library-based approach to synthesis of analog systems from VHDL-AMS specifications. ACM Trans. Design Autom. Electron. Syst. **9**, 238–271 (2004)
21. K. Matsukawa, T. Morie, Y. Tokunaga, S. Sakiyama, Y. Mitani, M. Takayama, T. Miki, A. Matsumoto, K. Obata, S. Dosho, Design methods for pipeline and delta-sigma A-to-D converters with convex optimization, in *Asia and South Pacific Design Automation Conference* (2009), pp. 690–695
22. L.W. Nagel, SPICE2: A computer program to simulate semiconductor circuits. PhD Thesis, EECS Department, University of California, Berkeley, 1975
23. W. Nye, D.C. Riley, A. Sangiovanni-Vincentelli, A.L. Tits, DELIGHT.SPICE: an optimization-based system for the design of integrated circuits. IEEE Trans. Comput. Aided Des. Integr. Circuits Syst. **7**, 501–519 (1988)
24. F. Medeiro, F.V. Fernández, R. Dominguez-Castro, A. Rodriguez-Vázquez, A statistical optimization-based approach for automated sizing of analog cells, in *IEEE/ACM International Conference on Computer-Aided Design* (1994), pp. 594–597
25. J.R. Koza, F.H. Bennett, D. Andre, M.A. Keane, F. Dunlap, Automated synthesis of analog electrical circuits by means of genetic programming. IEEE Trans. Evolut. Comput. **1**, 109–128 (1997)
26. R. Phelps, M. Krasnicki, R.A. Rutenbar, L.R. Carley, J.R. Hellums, ANACONDA: simulation-based synthesis of analog circuits via stochastic pattern search. IEEE Trans. Comput. Aided Des. Integr. Circuits Syst. **19**, 703–717 (2000)
27. M. Krasnicki, R. Phelps, R.A. Rutenbar, L.R. Carley, MAELSTROM: efficient simulation-based synthesis for custom analog cells, in *Design Automation Conference* (1999), pp. 945–950
28. G. Zhang, A. Dengi, L.R. Carley, Automatic synthesis of a 2.1 GHz SiGe low noise amplifier, in *IEEE Radio Frequency Integrated Circuits (RFICs) Symposium* (2002), pp. 125–128

29. G. Alpaydin, S. Balkir, G. Dundar, An evolutionary approach to automatic synthesis of high-performance analog integrated circuits. IEEE Trans. Evol. Comput. **7**, 240–252 (2003)
30. P. Vancorenland, C.D. Ranter, M. Steyaert, G. Gielen, Optimal RF design using smart evolutionary algorithms, in *Design Automation Conference* (2000), pp. 7–10
31. E. Malavasi, H. Chang, A. Sangiovanni-Vincentelli, E. Charbon, U. Choudhury, E. Felt, G. Jusuf, E. Liu, R. Neff, *A Top-Down, Constraint-Driven Design Methodology for Analog Integrated Circuits* (Springer, Berlin, 1993)
32. I. Vassiliou, Design methodologies for RF and mixed-signal systems. PhD Thesis, EECS Department, University of California, Berkeley, 1999
33. G.G.E. Gielen, System-level design tools for RF communication ICs, in *URSI International Symposium on Signals, Systems, and Electronics* (1998), pp. 422–426
34. R. Gonzalez-Echevarria, E. Roca, R. Castro-López, F.V. Fernández, J. Sieiro, J.M. López-Villegas, N. Vidal, An automated design methodology of RF circuits by using Pareto-optimal fronts of EM-simulated inductors. IEEE Trans. Comput. Aided Des. Integr. Circuits Syst. **36**, 15–26 (2017)
35. G. Gielen, T. Eeckelaert, E. Martens, T. McConaghy, Automated synthesis of complex analog circuits, in *European Conference on Circuit Theory and Design* (2007), pp. 20–23
36. R. Póvoa, I. Bastos, N. Lourenço, N. Horta, Automatic synthesis of RF front-end blocks using multi-objective evolutionary techniques. Integr. VLSI J. **52**, 243–252 (2016)
37. F. Medeiro, R. Rodríguez-Macías, F.V. Fernández, R. Domínguez-Castro, J.L. Huertas, A. Rodríguez-Vázquez, Global design of analog cells using statistical optimization techniques. Analog Integr. Circ. Sig. Process. **6**, 179–195 (1994)
38. J. Rijmenants, J.B. Litsios, T.R. Schwarz, M.G.R. Degrauwe, ILAC: an automated layout tool for analog CMOS circuits. IEEE J. Solid State Circ. **24**, 417–425 (1989)
39. E. Malavasi, E. Charbon, E. Felt, A. Sangiovanni-Vincentelli, Automation of IC layout with analog constraints. IEEE Trans. Comput. Aided Des. Integr. Circuits Syst. **15**, 923–942 (1996)
40. SpectreRF. https://www.cadence.com/content/cadence-www/global/en_US/home/tools/custom-ic-analog-rf-design/circuit-simulation/spectre-rf-option.html. Accessed 28 Jan 2020
41. EldoRF. https://www.mentor.com/tannereda/eldo-rf. Accessed 28 Jan 2020
42. HspiceRF. https://www.synopsys.com/verification/ams-verification/hspice/whats-new-in-hspice.html. Accessed 28 Jan 2020
43. N. Pulsford, Passive integration technology: targeting small accurate RF parts. RF Des. **25**, 40–48 (2002)
44. G. Tulunay, S. Balkir, Synthesis of RF CMOS low noise amplifiers, in *IEEE International Symposium on Circuits and Systems* (2008), pp. 880–883
45. C. De Ranter, G. Van der Plas, M. Steyaert, G. Gielen, W. Sansen, CYCLONE: automated design and layout of RF LC-oscillators. IEEE Trans. Comput. Aided Des. Integr. Circuits Syst. **21**, 1161–1170 (2002)
46. R. Gupta, B.M. Ballweber, D.J. Allstot, Design and optimization of CMOS RF power amplifiers. IEEE J. Solid State Circ. **36**, 166–175 (2001)
47. C. Yue, S. Wong, Physical modeling of spiral inductors on silicon. IEEE Trans. Electron Devices **47**, 560–568 (2000)
48. A. Nieuwoudt, T. Ragheb, Y. Massoud, Hierarchical optimization methodology for wideband low noise amplifiers, in *Asia and South Pacific Design Automation Conference* (2007), pp. 68–73
49. Y. Xu, K.L. Hsiung, X. Li, L.T. Pileggi, S.P. Boyd, Regular analog/RF integrated circuits design using optimization with recourse including ellipsoidal uncertainty. IEEE Trans. Comput. Aided Des. Integr. Circuits Syst. **28**, 623–637 (2009)
50. E. Afacan, G. Dundar, A mixed domain sizing approach for RF circuit synthesis, in *IEEE International Symposium on Design and Diagnostics of Electronic Circuits Systems* (2016), pp. 1–4
51. F. Passos, M. Kotti, R. González-Echevarría, M.H. Fino, M. Fakhfakh, E. Roca, R. Castro-López, F.V. Fernández, Physical vs. surrogate models of passive RF devices, in *IEEE International Symposium on Circuits and Systems* (2015), pp. 117–120

52. R. González-Echevarría, R. Castro-López, E. Roca, F.V. Fernández, J. Sieiro, N. Vidal, J. López-Villegas, Automated generation of the optimal performance trade-offs of integrated inductors. IEEE Trans. Comput. Aided Des. Integr. Circuits Syst. **33**, 1269–1273 (2014)
53. S.K. Mandal, S. Sural, A. Patra, ANN- and PSO-based synthesis of on-chip spiral inductors for RF ICs. IEEE Trans. Comput. Aided Des. Integr. Circuits Syst. **27**, 188–192 (2008)
54. B. Liu, D. Zhao, P. Reynaert, G.G.E. Gielen, Synthesis of integrated passive components for high-frequency RF ICs based on evolutionary computation and machine learning techniques. IEEE Trans. Comput. Aided Des. Integr. Circuits Syst. **30**, 1458–1468 (2011)
55. M. Ballicchia, S. Orcioni, Design and modeling of optimum quality spiral inductors with regularization and Debye approximation. IEEE Trans. Comput. Aided Des. Integr. Circuits Syst. **29**, 1669–1681 (2010)
56. I. Paenke, J. Branke, Y. Jin, Efficient search for robust solutions by means of evolutionary algorithms and fitness approximation, IEEE Trans. Evol. Comput. **10**, 405–420 (2006)
57. M.T.M. Emmerich, K.C. Giannakoglou, B. Naujoks, Single- and multiobjective evolutionary optimization assisted by Gaussian random field metamodels. IEEE Trans. Evol. Comput. **10**, 421–439 (2006)
58. J. Knowles, ParEGO: a hybrid algorithm with on-line landscape approximation for expensive multiobjective optimization problems. IEEE Trans. Evol. Comput. 10, 50–66 (2006)
59. Q. Zhang, W. Liu, E. Tsang, B. Virginas, Expensive multiobjective optimization by MOEA/D with Gaussian process model. IEEE Trans. Evol. Comput. **14**, 456–474 (2010)
60. F. Passos, E. Roca, R. Castro-López, F.V. Fernández, Radio-frequency inductor synthesis using evolutionary computation and Gaussian-process surrogate modeling. Appl. Soft Comput. **60**, 495–507 (2017)
61. M. Ranjan, A. Bhaduri, W. Verhaegen, B. Mukherjee, R. Vemuri, G. Gielen, A. Pacelli, Use of symbolic performance models in layout-inclusive RF low noise amplifier synthesis, in *IEEE International Behavioral Modeling and Simulation Conference* (2004), pp. 130–134
62. T. Liao, L. Zhang, Parasitic-aware GP-based many-objective sizing methodology for analog and RF integrated circuits, in *Asia and South Pacific Design Automation Conference* (2017), pp. 475–480
63. G. Zhang, A. Dengi, R.A. Rohrer, R.A. Rutenbar, L.R. Carley, A synthesis flow toward fast parasitic closure for radio-frequency integrated circuits, in *Design Automation Conference* (2004), pp. 155–158
64. A. Agarwal, R. Vemuri, Layout-aware RF circuit synthesis driven by worst case parasitic corners, in *International Conference on Computer Design* (2005), pp. 444–449
65. D. Ghai, S.P. Mohanty, E. Kougianos, Design of parasitic and process-variation aware nano-CMOS RF circuits: A VCO case study. IEEE Trans. Very Large Scale Integra. (VLSI) Syst. **17**, 1339–1342 (2009)
66. P. Vancorenland, G.V. der Plas, M. Steyaert, G. Gielen, W. Sansen, A layout-aware synthesis methodology for RF circuits, in *IEEE/ACM International Conference on Computer Aided Design* (2001), pp. 358–362
67. A. Bhaduri, V. Vijay, A. Agarwal, R. Vemuri, B. Mukherjee, P. Wang, A. Pacelli, Parasitic-aware synthesis of RF LNA circuits considering quasi-static extraction of inductors and interconnects, in *Midwest Symposium on Circuits and Systems*, vol. 1 (2004), pp. I–477–80
68. J. Kampe, C. Wisser, G. Scarbata, Module generators for a regular analog layout, in *International Conference on Computer Design VLSI in Computers and Processors* (1996), pp. 280–285
69. R. Castro-López, O. Guerra, E. Roca, F.V. Fernández, An integrated layout-synthesis approach for analog ICs. IEEE Trans. Comput. Aided Des. Integr. Circuits Syst. **27**, 1179–1189 (2008)
70. G. Berkol, A. Unutulmaz, E. Afacan, G. Dundar, F.V. Fernández, A.E. Pusane, F. Baskaya, A two-step layout-in-the-loop design automation tool, in *IEEE International New Circuits and Systems Conference* (2015), pp. 1–4
71. N. Lourenço, R. Martins, A. Canelas, R. Póvoa, N. Horta, AIDA: layout-aware analog circuit-level sizing with in-loop layout generation. Integra. VLSI J. **55**, 316–329 (2016)

72. N. Lourenço, R. Martins, N. Horta, Layout-aware sizing of analog ICs using floorplan & routing estimates for parasitic extraction, in *Design, Automation & Test in Europe Conference & Exhibition* (2015), pp. 1156–1161
73. T. McConaghy, K. Breen, J. Dyck, A. Gupta, *Variation-Aware Design* (Springer, Berlin, 2013), pp. 169–186
74. N. Lourenço, N. Horta, GENOM-POF: multi-objective evolutionary synthesis of analog ICs with corners validation, in *Annual Conference on Genetic and Evolutionary Computation* (2012), pp. 1119–1126
75. T. McConaghy, G.G.E. Gielen, Globally reliable variation-aware sizing of analog integrated circuits via response surfaces and structural homotopy. IEEE Trans. Comput. Aided Des. Integr. Circuits Syst. **28**, 1627–1640 (2009)
76. G. Berkol, E. Afacan, G. Dundar, A.E. Pusane, F. Başkaya, A novel yield aware multi-objective analog circuit optimization tool, in *2015 IEEE International Symposium on Circuits and Systems (ISCAS)* (2015), pp. 2652–2655
77. B. Liu, F.V. Fernández, G.G.E. Gielen, Efficient and accurate statistical analog yield optimization and variation-aware circuit sizing based on computational intelligence techniques. IEEE Trans. Comput. Aided Des. Integr. Circuits Syst. **30**, 793–805 (2011)
78. A. Canelas, R. Martins, R. Póvoa, N. Lourenço, N. Horta, Efficient yield optimization method using a variable K-means algorithm for analog IC sizing, in *Design, Automation & Test in Europe Conference & Exhibition* (2017), pp. 1201–1206
79. J. Crols, S. Donnay, M. Steyaert, G. Gielen, A high-level design and optimization tool for analog RF receiver front-ends, in *IEEE International Conference on Computer Aided Design* (1995), pp. 550–553
80. D. Gonzalez, A. Rusu, M. Ismail, Tackling 4G challenges with "TACT" - design and optimization of 4G radio receivers with a transceiver architecture comparison tool (TACT). IEEE Circ. Devices Mag. **22**, 16–23 (2006)
81. M. El-Nozahi, K. Entesari, E. Sanchez-Sinencio, A systematic system level design methodology for dual band CMOS RF receivers, in *Midwest Symposium on Circuits and Systems* (2007), pp. 1014–1017
82. S. Rodriguez, J.G. Atallah, A. Rusu, L.-R. Zheng, M. Ismail, ARCHER: an automated RF-IC Rx front-end circuit design tool. Analog Integr. Circ. Sig. Process. **58**, 255–270 (2009)
83. Z. Pan, C. Qin, Z. Ye, Y. Wang, Automatic design for analog/RF front-end system in 802.11ac receiver, in *Asia and South Pacific Design Automation Conference* (2015), pp. 454–459
84. B. Liu, G.G.E. Gielen, F.V. Fernández, *Automated Design of Analog and High-frequency Circuits - A Computational Intelligence Approach, Studies in Computational Intelligence* (Springer, Berlin, 2014)
85. G. Tulunay, S. Balkir, A synthesis tool for CMOS RF low-noise amplifiers. IEEE Trans. Comput. Aided Des. Integr. Circuits Syst. **27**, 977–982 (2008)
86. M. Chu, D.J. Allstot, Elitist nondominated sorting genetic algorithm based RF IC optimizer. IEEE Trans. Circ. Syst. I, Reg. Papers **52**, 535–545 (2005)

Chapter 2
RF Receiver Architectures

The objective of a communication system is to send information from a transmitter to a receiver. Therefore, the transmitter has to be appropriately designed in order to send signals with adequate power levels, and the receiver has to be able to receive a signal and extract the message. In this book we will focus on the receiver part of the communication chain. Therefore, this chapter will present a brief overview of this system, providing the main performances and parameters for each block included.

The goal of any radio receiver is to extract and selectively detect a desired signal from the complete electromagnetic spectrum. The ability to select a given signal in the presence of the huge amount of interfering signals and noise is the fundamental task for radio receivers. Nowadays, radio receivers must often be able to detect signal powers as small as a few nano-watts, while rejecting a multitude of other signals that may be several orders of magnitude larger [1]. Because the electromagnetic spectrum is a limited resource, interfering signals often lie very close to the desired signal, thereby increasing the difficulty of rejecting the unwanted signals. Over the times, different receiver architectures were proposed, each one with its pros and cons. Since in this book the final objective is to achieve the design of an RF front-end, it is important to review the different available receiver architectures. Hence, in Sect. 2.1 a brief description of some of the most common radio architectures is performed. Afterwards, in Sect. 2.2, the blocks that constitute the RF front-end, and that will be designed throughout this book are reviewed. Their main purpose at the receiver chain is discussed and their most important performance parameters are examined.

© Springer Nature Switzerland AG 2020
F. Passos et al., *Automated Hierarchical Synthesis of Radio-Frequency Integrated Circuits and Systems*, https://doi.org/10.1007/978-3-030-47247-4_2

2.1   Receiver Architectures

2.1.1   Superheterodyne Receiver

The superheterodyne receiver architecture was the dominant choice for many decades [2]. Its generic architecture is shown in Fig. 2.1.

The superheterodyne features a band-selection filter between the antenna and the LNA, which rejects the out-of-band interference. Then, the signal is amplified by an LNA which is then followed by an image-rejection filter. This filter has the purpose of rejecting the unwanted image frequency band (which will be explained later). A mixer then converts the RF signal to a lower-frequency, which is commonly referred to as intermediate frequency (IF). Both the RF signal and a local oscillator (LO) signal (which can be generated by a VCO) enter the mixer, thereby generating the IF signal that appears at the mixers' output. The frequency of this IF signal is equal to the difference between the RF input signal frequency and the LO signal frequency. Therefore, the channel selection is performed by changing the oscillation frequency of the VCO. After the frequency down-conversion operation, a channel filter is implemented in the IF stage to remove any unwanted signals. Next, an automatic gain control (AGC) amplifier provides a significant amount of gain to the IF signal. The amplified IF signal is then demodulated, allowing the information to be processed. This architecture is known for being able to select narrow band signals from an environment full of interferers [3]. Since the channel selection is carried out at IF the dynamic range requirements are relaxed at baseband, which simplifies the design of the analog-to-digital converter [2].

One of the disadvantages of superheterodyne receivers is the image problem [3]. To understand this issue, note that an analog multiplier does not preserve the polarity of the difference between two signals, i.e., for $x_1(t) = A_1\cos(\omega_1 t)$ and $x_2(t) = A_2\cos(\omega_2 t)$, the product of $x_1(t)$ and $x_2(t)$ signals is $\cos[(\omega_1 - \omega_2)t]$, which is not different from $\cos[(\omega_2 - \omega_1)t]$. Therefore, in the superheterodyne architecture, the bands symmetrically located above and below the LO frequency are down-converted to the same center frequency at IF (see Fig. 2.2). If the band of interest is f_1, then the image is around $2f_{LO} - f_1$ and vice versa. Due to this problem, the noise present at the image band is also translated into the desired band.

Therefore, the total noise at IF is composed by the noise at the desired RF band, down-converted to IF, plus the noise at the image RF band up-converted to

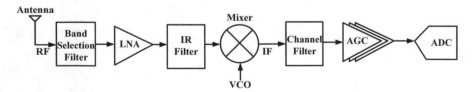

Fig. 2.1 Typical superheterodyne integrated RF receiver signal chain

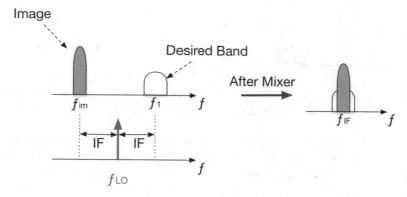

Fig. 2.2 Illustrating the image problem in superheterodyne receivers

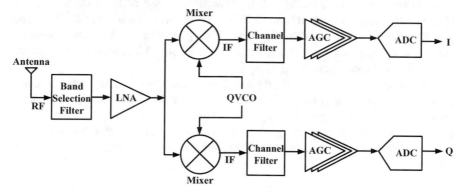

Fig. 2.3 Typical zero-IF integrated RF receiver signal chain

IF and also the noise added by the mixer circuit itself. Therefore, in this type of architecture, high quality factor filtering is usually required in order to meet the specifications of the communication standards. In order to design such high quality factor filters, the image rejection (IR) filter and/or the channel selection IF filter have to be implemented off-chip. Also, different image rejection and channel selection filters have to be used for different standards, making it difficult to achieve a highly integrated low cost solution [4]. Therefore, this receiver is not particularly suited for IC design.

2.1.2   Zero-IF Receiver

A receiver with its IF set to zero is called zero-IF, direct conversion or homodyne receiver. The architecture of this type of receiver is shown in Fig. 2.3.

It can be seen in Fig. 2.3 that a Quadrature-VCO (QVCO) drives two different mixers. The QVCO generates two different signals with opposite phases which then drive two different mixers. It is shown in [3] that the signals at the output of the mixers have the same polarity, containing their images at opposite polarity. Thus, in the ideal case, the sum of the two output signals is an image-free signal. This is the main advantage of zero-IF receivers. Therefore, this architecture is more suitable for integration than the standard superheterodyne receiver because no IR filter is needed.

However, some disadvantages outcome from the use of this architecture. The main disadvantage of these receivers is the appearance of a DC offset at the mixer output. Since in this architecture the down-converted band is at zero frequency, inappropriate offset voltages can corrupt the signal and saturate the following stages. The causes for the DC offset can be observed in Fig. 2.4. The first cause for DC offset is the so-called LO leakage. This phenomenon appears due to capacitive and substrate coupling. The signal from the LO now appears at the input of the LNA and the mixer signal is *mixed* with this signal. This phenomenon is sometimes called "self-mixing" [3]. A similar effect occurs if a large signal interferer leaks from the LNA to the LO and is multiplied by itself.

Another problem of zero-IF architectures is the Flicker noise. Flicker noise (sometimes also referred to as $1/f$ noise) appears in MOS transistors when these are operating near the zero frequency range. This noise is usually generated when a direct current flows through the MOS transistors. Since in zero-IF receivers we are translating the signal to zero frequency, this noise usually highly affects this type of receivers.

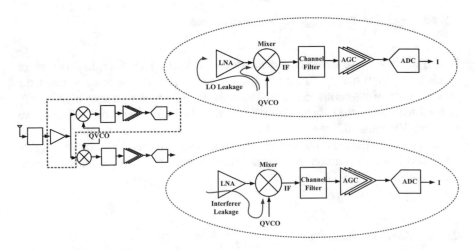

Fig. 2.4 Self-mixing of the LO signal and a strong interferer

2.1.3   Low-IF Receiver

The low-IF receiver architecture has a block diagram similar to the zero-IF one shown in Fig. 2.3. Low-IF receivers perform frequency down-conversion of the signal to frequencies close to, but higher than zero. This architecture presents a trade-off between the superheterodyne and the zero-IF receivers. If on the one hand, the DC offset and Flicker noise problems of the zero-IF receivers are solved, on the other hand, the image problem is back and this time with the additional difficulty of filtering it at a very low frequency.

There are several other receiver architectures available in literature [2, 5]. Depending on the application, each architecture presents its pros and cons, and the designer must choose it intelligently and carefully. In the next section, the receiver blocks that constitute the front-end receivers are briefly described, as well as its main performance parameters.

2.2   Front-End Receiver Blocks

The three main blocks of the front-end receiver, which are the common to all receiver architectures are: the LNA, the VCO, and the mixer. In this section these blocks are introduced and their most important performances are described. Nevertheless, there are two main performances that are common not only to these three blocks but also to all CMOS circuits in general, which are the power consumption and the area occupation.

2.2.1   Power Consumption

Power is very important in RF circuit design. Nowadays, in an era where all devices aim at portability and at continuous interconnection (e.g., IoT), the battery of a given device is very important. Therefore, power consumption has to be kept at a minimum. It is possible to understand that higher power dissipation limits a circuit because portable electronic devices are limited by battery life. Therefore, power is one of the most important system performance metrics. When we speak about the power consumption of a circuit we are usually interested in the DC power consumption given by

$$P_{DC} = V_{DD} \cdot I_{DC} \tag{2.1}$$

where V_{DD} is the power supply voltage and I_{DC} is the total DC current of the circuit. Therefore, if the designer wishes to minimize the power consumption, he/she can reduce power supply levels or reduce the current level, or, if possible, reduce both.

2.2.2   Area Occupation

The area in CMOS integrated circuits is directly associated with cost because larger area reduces the number of dies that can fit into a wafer and therefore leads to a linear increase in processing and material costs. Another significant issue is the impact of die area on die yield. It has been reported that the yield (the fraction of the fabricated ICs that are fully functional) decreases sharply with die area. As a result, die manufacturing costs quickly become prohibitive beyond some size determined by the process technology not only by the cost associated with area, but also due to the yield of a given die [6].

2.2.3   Low Noise Amplifier

An LNA is an electronic circuit whose objective is to amplify a very low-power signal without significantly degrading its signal-to-noise ratio (SNR). The SNR is a measure used to compare the level of a desired signal strength to the level of noise (see Fig. 2.5).

The LNA symbol is shown in Fig. 2.6. The LNA is usually one of the first components just after the antenna in an integrated wireless receiver chain. The weak signal received at the antenna is fed into the LNA. In order to absorb as much signal power as possible from the antenna, the LNA needs to provide a sufficiently matched input impedance. Being one of the first elements in the receiver chain the LNA is a key block for the noise performance of the whole chip.

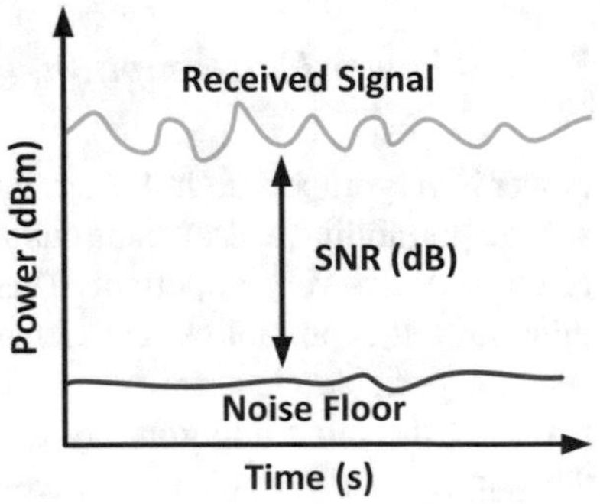

Fig. 2.5 Illustrating the SNR

Fig. 2.6 Low noise amplifier symbol

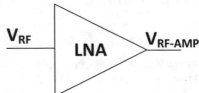

Fig. 2.7 Two-port network

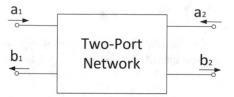

The LNA can be considered as a two-port system and, therefore, it is important to establish the concept of Scattering Parameters, or S-Parameters for short. The S-parameters describe the response of an N-port network to incident signals, to any or all of the ports. Consider, for instance, the two-port network illustrated in Fig. 2.7.

The signal at each of the ports can be thought of as the superposition of two waves traveling in opposite directions (e.g., a_1 and b_1). By convention, each port is shown as two nodes in order to give a name and value to these opposite direction waves. The variable a_i represents a wave incident to port i and the variable b_j represents a wave reflected from port j. The magnitude of the a_i and b_j variables can be thought of as voltage-like variables, normalized using a specified reference impedance (e.g., Z_0). The S-parameters can be defined as,

$$S_{11} = \frac{b_1}{a_1}$$

$$S_{12} = \frac{b_1}{a_2}$$

$$S_{21} = \frac{b_2}{a_1}$$

(2.2)

$$S_{22} = \frac{b_2}{a_2}$$

where S_{11} is the input reflection coefficient, S_{22} is the output reflection coefficient, S_{21} is the forward transmission coefficient, and S_{12} is the reverse forward transmission coefficient. The first number in the S-Parameter subscript refers to the responding port, while the second number refers to the incident port. Hence, S_{21} means the response at port 2 due to a signal at port 1. These S-parameters are important for the understanding of the LNA performance parameters, which are described below.

2.2.3.1 Gain

The gain is one of the most important performance measures of LNAs. The gain quantifies the ability of a system to increase the amplitude of an input signal and in LNAs it is given by the forward transmission coefficient, S_{21}.

2.2.3.2   Noise Figure

Noise figure (NF) is a measure of degradation of the signal-to-noise (SNR) ratio in a given circuit. The NF can be defined as

$$NF = 10 \cdot \log_{10} \left(\frac{SNR_{in}}{SNR_{out}} \right) \tag{2.3}$$

where SNR_{in} and SNR_{out} are the signal-to-noise ratio measured at the input and output of the circuit, respectively.

2.2.3.3   Linearity

The basic idea for the calculation of non-linear effects is that the behavior of a circuit can be approximated by a non-linear equation:

$$y(t) = a_1 \cdot x(t) + a_2 \cdot x2(t) + a_3 \cdot x^3(t) + \ldots \tag{2.4}$$

where

- $y(t)$ is the output signal
- $x(t)$ is the input signal
- $a_1 > 0$
- $a_2 > 0$
- $a_3 < 0$

If the circuit is considered ideally balanced, then the even order components are set to zero e.g., $a_2 = 0$ for convenience. When the input signal is a single-tone $x(t) = A \cdot \cos(\omega_0 t)$, the generated output signal $y(t)$ consists of different frequency components:

$$
\begin{aligned}
y(t) = {} & \frac{1}{2} a_2 A^2 && \text{// corresponds to } dc \text{ component} \\[2mm]
& + \left(a_1 A + \frac{3}{4} a_3 A^3 \right) \cos(\omega_0 t) && \text{// corresponds to } \omega_0 \text{ component} \\[2mm]
& + \frac{1}{2} a_2 A^2 \cos(2\omega_0 t) && \text{// corresponds to } 2\omega_0 \text{ component} \\[2mm]
& + \frac{1}{4} a_3 A^3 \cos(3\omega_0 t) && \text{// corresponds to } 3\omega_0 \text{ component} \\[2mm]
& + \ldots
\end{aligned}
\tag{2.5}
$$

Besides the emerging spectral components, it is worth noting that the linear amplification of the output fundamental is no longer a_1 but $(a_1 + \frac{3}{4} a_3 A^2)$. This means that the third-order non-linearity reduces the voltage gain of the fundamental sine wave. The phenomenon is well known in literature and commonly characterized by the 1 dB compression point (CP_1) [3]. The CP_1 is defined as the power (or voltage) level for which the amplification of the fundamental frequency is attenuated by 1 dB compared to its ideal linear amplification a_1. From (2.5) it can be shown that the input-referred 1 dB compression point (CP_{1i}) is reached for

$$A_{1\,dB} = \sqrt{0.145|a_1/a_3|} \tag{2.6}$$

Let us assume an input signal consisting of two fundamental tones $\omega = \omega_1$ and $\omega = \omega_2$ with different amplitudes A_1 and A_2:

$$x(t) = A_1\cos(\omega_1 t) + A_2\cos(\omega_2 t) \tag{2.7}$$

The output signal $y(t)$ results in

$$
\begin{aligned}
y(t) \;=\; & \tfrac{a_2}{2}\left(A_1^2 + A_2^2\right) && // \quad \text{corresponds to } dc \text{ component}\\
& + \left(a_1 + a_3\left(\tfrac{3}{4}A_1^2 + \tfrac{3}{2}A_2^2\right)\right)\\
& \times A_1\cos(\omega_1 t) && // \quad \text{corresponds to } \omega_1 \text{ component}\\
& + \left(a_1 + a_3\left(\tfrac{3}{4}A_2^2 + \tfrac{3}{2}A_1^2\right)\right)\\
& \times A_2\cos(\omega_2 t) && // \quad \text{corresponds to } \omega_2 \text{ component}\\
& + \tfrac{a_2}{2}A_1^2\cos(2\omega_1 t) && // \quad \text{corresponds to } 2\omega_1 \text{ component}\\
& + \tfrac{a_2}{2}A_2^2\cos(2\omega_2 t) && // \quad \text{corresponds to } 2\omega_2 \text{ component}\\
& + a_2 A_1 A_2\cos((\omega_1 + \omega_2)t) && // \quad \text{corresponds to } \omega_1 + \omega_2 \text{ component}\\
& + a_2 A_1 A_2\cos((\omega_1 - \omega_2)t) && // \quad \text{corresponds to } \omega_1 - \omega_2 \text{ component}\\
& + \tfrac{a_3}{4}A_1^3\cos(3\omega_1 t) && // \quad \text{corresponds to } 3\omega_1 \text{ component}\\
& + \tfrac{a_3}{4}A_2^3\cos(3\omega_2 t) && // \quad \text{corresponds to } 3\omega_2 \text{ component}\\
& + \tfrac{3}{4}a_3 A_1^2 A_2\cos((2\omega_1 + \omega_2)t) && // \quad \text{corresponds to } 2\omega_1 + \omega_2 \text{ component}\\
& + \tfrac{3}{4}a_3 A_1 A_2^2\cos((\omega_1 + 2\omega_2)t) && // \quad \text{corresponds to } \omega_1 + 2\omega_2 \text{ component}\\
& + \tfrac{3}{4}a_3 A_1^2 A_2\cos((2\omega_1 - \omega_2)t) && // \quad \text{corresponds to } 2\omega_1 - \omega_2 \text{ component}\\
& + \tfrac{3}{4}a_3 A_1 A_2^2\cos((2\omega_2 - \omega_1)t) && // \quad \text{corresponds to } 2\omega_2 - \omega_1 \text{ component}\\
& + \ldots
\end{aligned}
\tag{2.8}
$$

First, let us assume that the input signals significantly differ in amplitude, i.e., $A_2 \gg A_1$. From (2.8) we see that the gain for the fundamental tone ω_2 will be reduced (desensitization) or may even drop to zero (blocking) [3]. If we assume that the second tone in (2.5) is modulated in amplitude by a sinusoid with $\omega = \omega_M$ e.g., $A_2(1 + m\cos(\omega_M t))$, where $m < 1$ is the modulation index and $A_2 \gg A_1$,

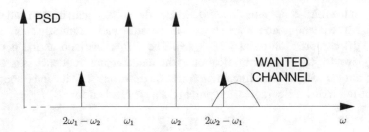

Fig. 2.8 Illustrating the third-order intermodulation product at $\omega = 2\omega_2 - \omega_1$, which falls into the wanted channel

(2.8) changes into,

$$y(t)$$

$$= \left[a_1 + \frac{2}{3}a_3 A_2^2 \left(1 + \frac{m^2}{2} + \frac{m^2}{2}\cos(2\omega_M t) + 2m\cos(\omega_M t) \right) \right] A_1\cos(\omega_1 t) + \dots$$

$$(2.9)$$

The gain for the signal at $\omega = \omega_1$ now contains amplitude modulation with $\omega = \omega_M$ and $\omega = 2\omega_M$. This phenomenon is called cross modulation [3]. Another interesting scenario is the presence of spectral components at $\omega = 2\omega_1 - \omega_2$ or $\omega = 2\omega_2 - \omega_1$ in the output signal $y(t)$ in (2.8). In Fig. 2.8, it is possible to observe two interferers at $\omega = \omega_1$ and $\omega = \omega_2$, which are located close to the wanted channel. The third-order intermodulation product at $\omega = 2\omega_2 - \omega_1$ falls into the wanted channel.

Assume two strong interferers with amplitudes A_1 and A_2 with a frequency spacing $\Delta\omega = \omega_2 - \omega_1$ so that the third-order intermodulation product occurring at $\omega = 2\omega_2 - \omega_1$ falls into a wanted frequency channel. If the wanted signal is too weak, the intermodulation product may dominate the spectra of the channel. Due to its position on the frequency axis, the intermodulation product cannot be filtered out. Literature commonly describes the effects of third-order intermodulation in terms of the third-order intercept point (IIP_3) (see Fig. 2.9). From (2.8) we see that the spectral power of the third-order products grows with the power of three while the fundamental components grow linearly. If we plot the output power P_{OUT} for the fundamental component and the third-order component versus the input power P_{IN} of the fundamental input tone of a non-linear circuit in logarithmic scale, it is possible to observe that the slope of the fundamental output tone is one while the slope of the third-order component is three (for small input powers far from the power levels where the circuit is already in compression). The parameter IIP_3 is commonly defined as the intersection of both lines.

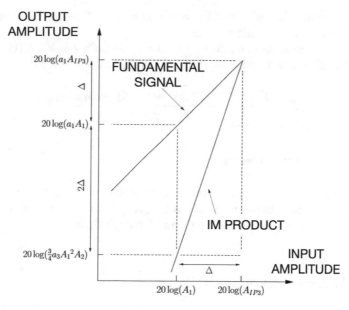

Fig. 2.9 Graphical interpretation of the IIP_3 in double logarithmic scaling. The intermodulation product is growing with the power of three while the fundamental signal is growing linearly

2.2.3.4 Input and Output Matching

The input and output matching are given by S_{11} (input matching) and S_{22} (output matching). When a circuit is connected to another, usually the input and output impedance do not entirely match. This means that part of the signal is not passed between circuits and is reflected back, losing efficiency in the process. This is especially important for the LNA, since its input is connected to the antenna. The antenna usually has a characteristic impedance of $50\,\Omega$ and since the LNA is intended to capture a weak signal, the receiver cannot afford to lose even more power, so the LNA must be carefully designed to match the antenna impedance.

However, while it is absolutely necessary to take care of input impedance matching for the LNA, where the incident and reflected power in the circuit really exists, in some specific cases of IC design, the designer does not need to have perfectly matched circuits since in an IC circuit the size of the die is so small that the incident and reflective power or voltage waves are redundant or meaningless [7].

2.2.3.5 Stability

When discussing amplifiers, stability refers to an amplifiers' immunity to oscillation. Stability can be conditional or unconditional. Conditional stability means that the design is stable at certain input/output impedances, however, there is a region, of either source or load impedances, that can cause the circuit to oscillate.

Unconditional stability refers to a network that can "see" any possible impedance and the design will not oscillate.

In order to evaluate the LNA stability, the Rollet's stability factor, K, can be used. This can be calculated from the S-parameters of the LNA as follows:

$$K = \frac{1 - |S_{11}|^2 - |S_{22}|^2 + |S_{11}S_{22} - S_{12}S_{21}|^2}{2|S_{12}||S_{21}|} \tag{2.10}$$

2.2.3.6 Port-to-Port Isolation

The port-to-port isolation is a measure of how well the input and output of the LNA are separated in terms of unwanted signal coupling. For the LNA this performance is given by S_{12}, the reverse forward transmission coefficient.

2.2.4 Voltage Controlled Oscillator

The VCO is a circuit whose oscillation frequency is controlled by an input voltage, referred to as tuning voltage. By changing this voltage, the oscillation frequency shifts. Therefore, the VCO is used in receivers in order to, e.g., tune into a communication channel. The VCO is usually illustrated with the symbol shown in Fig. 2.10.

The most important performance parameters in VCO designs are given below.

2.2.4.1 Oscillation Frequency

It is easy to understand that when discussing VCOs one of the most important parameters is the oscillation frequency, f_{osc}, which is the signal frequency generated at the output of the VCO.

Fig. 2.10 Voltage controlled oscillator symbol

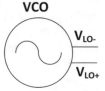

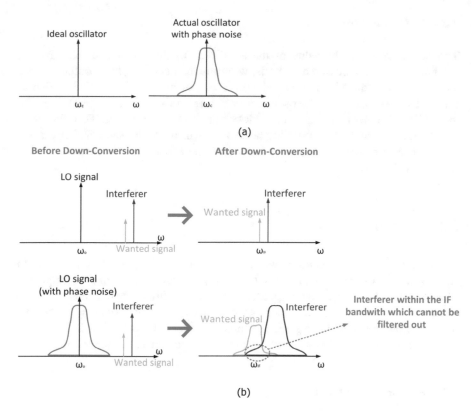

Fig. 2.11 (**a**) Illustrating the meaning of phase noise and (**b**) Phase noise problem in down-conversion receivers

2.2.4.2 Phase Noise

The phase noise (PN) is one of the main metrics of an oscillator in RF applications. In Fig. 2.11a it is possible to observe an ideal oscillator output in the frequency domain versus a real oscillator output. In the real situation, the signal spectrum shows symmetrical tails decreasing as $1/\omega_m^2$, where ω_m is the (angular) frequency offset from ω_0, which are referred to as the oscillator phase noise. This $1/\omega_m^2$ dependence suggests that a sort of white noise is appearing in the circuit which affects the signal integrity [8]. Figure 2.11b illustrates what happens when a VCO is used in a receiver for down-conversion, first in an ideal case where the LO signal does not have phase noise, and then, in a real case where the LO signal has phase noise. It can be seen in the latter that the down-converted signals have overlapping spectra with the wanted signal, after the down-conversion operation, suffering from significant noise contribution due to the tail of the interferer.

2.2.4.3   Output Swing

The output swing is the value of the amplitude of the signal generated by the oscillator. It can be measured in Volts, which is then called voltage output swing, V_{OUT}, or in dBm, which is called power output swing, P_{OUT}. This value is especially important when the oscillator polarizes a mixer (e.g., receiver). Usually, the VCO will polarize the gate of a MOS transistor of the mixer. Therefore, this value should be as high as possible because it will make the switching operation of the MOS transistors of the mixer faster and, therefore, it will decrease the noise introduced in the circuit.

2.2.5   Mixer

The mixer is a circuit which receives two different input signals (the RF and LO signals) at two different frequencies and outputs a signal at a different frequency (the so-called IF signal). This output frequency is the difference between the input frequencies, therefore, the designer can translate any input frequency to any IF frequency by tuning the LO signal. The mixer symbol is shown in Fig. 2.12.

The most important performance parameters for the mixer are the following.

2.2.5.1   Conversion Gain

As for the LNAs, the gain quantifies the ability of the mixer to increase the amplitude of the input signal. However, for the mixers the gain is referred to as conversion gain CG because the input and output signal are not at the same frequency.

2.2.5.2   Linearity

The linearity is also an important metric for the mixer. Since it was already introduced for the LNA, no additional discussion is required here.

Fig. 2.12 Mixer symbol

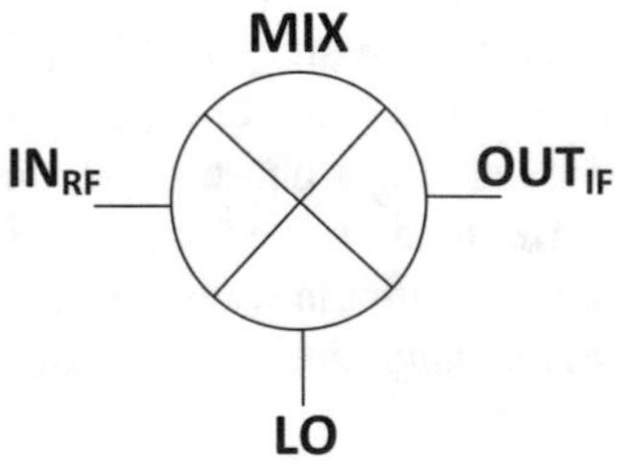

MIX

IN_{RF}

OUT_{IF}

LO

2.2.5.3   Port-to-Port Isolation

Similarly to the LNA, the mixer port-to-port isolation is a measure of how well the mixer ports are separated from each other in terms of unwanted signal coupling. This is very important because the LO-RF feedthrough results in LO leakage to the LNA, whereas RF-LO feedthrough allows strong interferer in the RF path to interact with the LO and driving the mixer. The LO-IF feedthrough is also very important because if a relatively high feedthrough exists from the LO port to the IF, the following stage could be desensitized. Desensitization is a phenomenon that may happen in receivers, where a strong interferer is present at the output of the receiver and, therefore, the receiver cannot identify the wanted signal [3]. The required isolation levels depend on the environment, but typically, an isolation of 30 dB is in most cases considered "high isolation" [3].

2.2.5.4   Noise Figure

The mixer noise figure analysis deserves special attention. As previously mentioned, all mixers fold the RF spectrum around the LO frequency, creating an output that contains the summation of the spectrum on both sides. In low-IF architectures, one of these contributions is typically considered spurious and the other intended. Therefore, image reject filtering is used to largely remove one of these responses. On the other hand, in zero-IF receivers, the case is different, because both sidebands are converted and used for the wanted signal. Due to this fact, different noise definitions arise: the single-sideband (SSB) noise and the double-sideband (DSB) noise figure.

The single-sideband noise figure (NF_{SSB}) assumes that the noise from both sidebands is folded into the output signal. However, since only one of the sidebands is useful for conveying the wanted signal, the other is filtered. Therefore, the noise level is doubled, without doubling the signal level, which naturally results in a 3 dB increase in noise figure (see Fig. 2.13).

The single-sideband (SSB) noise figure, which characterizes a low-IF receiver, is given by

$$NF_{SSB} = 10\log_{10}\left(\frac{S_{RF}/N_{RF+IM}}{(S/N)_{IF}}\right) \tag{2.11}$$

where S_{RF} is the signal power at the RF frequency, N_{RF+IM} is the noise power contributions from the RF and the image (IM) frequency and $(S/N)_{IF}$ is the signal-to-noise ratio at the intermediate frequency (IF).

On the other hand, the double-sideband noise figure (NF_{DSB}) assumes that both responses of the mixer contain parts of the wanted signal and, therefore, the noise is folded alongside with the corresponding signal. Therefore, the total NF is not impacted (see Fig. 2.14), because the signal is also translated. The double-sideband (DSB) noise figure characterizes a zero-IF receiver, and is given by

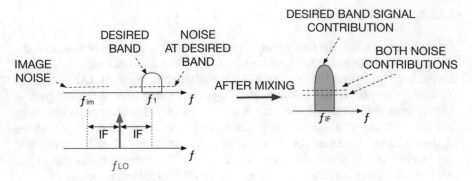

Fig. 2.13 Illustrating the NF_{SSB} in mixers

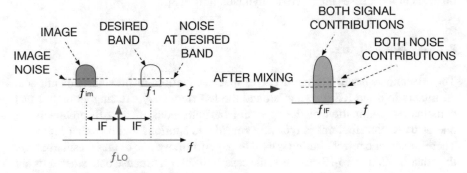

Fig. 2.14 Illustrating the NF_{DSB} in mixers

$$NF_{DSB} = 10\log_{10}\left(\frac{S_{RF+IM}/N_{RF+IM}}{(S/N)_{IF}}\right) \qquad (2.12)$$

where S_{RF+IM} is the signal power contribution at the RF and the IM frequency.

2.3   Summary

In this chapter different receiver architectures have been presented, as well as some of their advantages and disadvantages. The main blocks that constitute a receiver front-end have been explained (i.e., the LNA, the VCO, and the mixer) and also their performance parameters have been illustrated.

References

1. D.K. Shaeffer, *The Design and Implementation of Low-Power CMOS Radio Receivers* (Springer Science & Business Media, Berlin, 1999)
2. P.I. Mak, S.P. U, R.P. Martins, Transceiver architecture selection: review, state-of-the-art survey and case study. IEEE Circ. Syst. Mag. **7**(2), 6–25 (2007)
3. B. Razavi, *RF Microelectronics* (Prentice-Hall, Upper Saddle River, 1998)
4. D. Gonzalez, A. Rusu, M. Ismail, Tackling 4G challenges with "TACT" - design and optimization of 4G radio receivers with a transceiver architecture comparison tool (TACT). IEEE Circ. Device. Mag. **22**, 16–23 (2006)
5. S. Bronckers, A. Roc'h, B. Smolders, Wireless receiver architectures towards 5G: where are we? IEEE Circ. Syst. Mag. **17**(3), 6–16 (2017)
6. C. Toumazou, G. Moschytz, B. Gilbert, G. Kathiresan, *Trade-Offs in Analog Circuit Design*, 1st edn. (Kluwer Academic Publishers, Dordrecht, 2002)
7. R. Li, *RF Circuit Design*, 2nd edn. (Wiley, Hoboken, 2002)
8. C. Samori, Understanding phase noise in LC VCOs: a key problem in RF integrated circuits. IEEE Solid State Circ. Mag. **8**(4), 81–91 (2016)

Chapter 3
Modeling and Synthesis of Radio-Frequency Integrated Inductors

This chapter discusses the modeling and synthesis of integrated inductors. In Sect. 3.1, typical inductor topologies are shown, as well as their geometric and performance parameters. Furthermore, the inductor typical behavior over frequency is illustrated in order to give some design insights for this passive component. It is important to know the geometric and performance parameters of this device in order to understand how can the device be modeled and designed. Afterwards, in Sect. 3.2, two different inductor models are presented: a physical model, which relates analytical equations to a set of electrical components, which are able to emulate the behavior of the inductor, and, a surrogate model, based on machine learning techniques, which is able to capture the behavior of the inductor by learning from a set of *training* samples, and, afterwards, estimate its performances. Furthermore, in the same section, an accuracy comparison is performed between such physical and surrogate models. In Sect. 3.3 different optimization algorithms, both single- and multi-objective are used in order to synthesize integrated inductors. Some comparisons will be performed in order to find out which are the best optimization strategies to synthesize inductors in terms of accuracy and efficiency. Finally, Sect. 3.4 presents a tool that was developed incorporating the surrogate models presented in Sect. 3.2 and that has the ability to design and optimize several inductor topologies using single- and multi-objective optimization algorithms. Furthermore, the tool also allows the user to build new inductor models for new inductor topologies and technologies.

3.1 Integrated Inductor Design Insights

Inductors in RF integrated circuits are typically built by using two metal layers, with an intermediate dielectric layer. In Fig. 3.1, the shapes of octagonal asymmetric and symmetric spiral inductors are illustrated. The geometry of these inductors is usually

© Springer Nature Switzerland AG 2020
F. Passos et al., *Automated Hierarchical Synthesis of Radio-Frequency Integrated Circuits and Systems*, https://doi.org/10.1007/978-3-030-47247-4_3

Fig. 3.1 Different inductor topologies. (**a**) Octagonal asymmetric inductor topology and (**b**) octagonal symmetric inductor topology

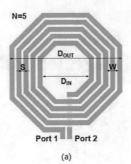 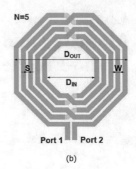

(a)   (b)

defined by four geometric parameters: number of turns (N), inner diameter (D_{in}) (or alternatively the outer diameter (D_{out})), turn width (w), and spacing between turns (s).

The most relevant inductor performances are the equivalent inductance, L_{eq}, and the quality factor, Q_{eq}. These performance parameters can be easily obtained from the S-parameters of the two-port structure representation of the inductor. For the asymmetric topology, the inductor is measured in single ended mode (SEM), and, therefore, the following formulas can be used in order to calculate L_{eq}, and Q_{eq} [1],

$$S_{SEM} = S_{11} - \frac{S_{12} \cdot S_{21}}{1 + S_{22}}$$

$$Z_{eq} = Z_0 \cdot \frac{1 + S_{SEM}}{1 - S_{SEM}} \tag{3.1}$$

where Z_{eq} is the equivalent input impedance and Z_0 the characteristic impedance of the device (usually 50 Ω). For the symmetric topology, the inductor is measured in differential mode (DM) and, the following formulas can be used [1],

$$S_{DM} = \frac{S_{11} - S_{12} - S_{21} + S_{22}}{2}$$

$$Z_{eq} = 2 \cdot Z_0 \cdot \frac{1 + S_{DM}}{1 - S_{DM}} \tag{3.2}$$

After obtaining the equivalent input impedance, Z_{eq}, (either for the asymmetric or symmetric inductor), L_{eq}, and the quality factor, Q_{eq}, are calculated as follows:

$$L_{eq}(f) = \frac{Im[Z_{eq}(f)]}{2\pi f} \tag{3.3}$$

$$Q_{eq}(f) = \frac{Im[Z_{eq}(f)]}{Re[Z_{eq}(f)]} \tag{3.4}$$

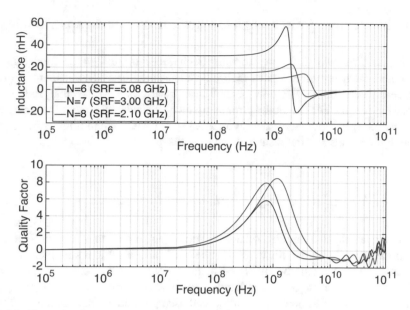

Fig. 3.2 Illustrating inductance and quality factor as a function of frequency for three different inductors

where f stands for frequency. In Fig. 3.2, three different plots of the inductance and quality factor as a function of the frequency are illustrated. An important parameter is the self-resonance frequency, SRF, which is defined as the frequency at which the imaginary part of Z_{eq} is zero, or, equivalently, the frequency at which the behavior of the inductor changes from inductive to capacitive (see Eq. (3.3)). While designing an inductor, the designer is usually interested in obtaining the inductance at the operation frequency and its quality factor. It is also important to notice that inductors represent a large percentage of the RF circuit area, and, therefore, they should be designed with the smallest area occupation since fabrication cost grows linearly with area. The area can be immediately calculated from the geometric parameters and does not need a model for its estimation.

After presenting the geometric and performance parameters of integrated inductors, in the following section, two different techniques will be studied in order to model these passive components and estimate their performance parameters. The basic idea of the modeling techniques is to estimate the inductors performances from its geometric parameters. Both models will be compared in terms of accuracy and efficiency.

3.2   Modeling Methodologies: Physical vs. Surrogate Models

In this section, two different modeling techniques for integrated inductors are presented. Both techniques have their advantages and disadvantages, and provide different trade-offs to RF designers. These trade-offs will be analyzed in the following section. For the model error assessment, a statistical study of an octagonal asymmetric spiral inductor topology in a $0.35\,\mu m$ CMOS technology is performed using both models and the results are compared.

3.2.1   Physical Modeling of Integrated Inductors

An inductor physical model relates a set of equations to a set of electrical components. Afterwards, this model can be used in an electrical simulator in order to emulate the inductor behavior. Also, the impedance Z_{eq} can be analytically calculated in order to get the inductor performances using Eqs. (3.3) and (3.4). One of the most well-known and widespread used physical models, is the π-model, illustrated in Fig. 3.3 and presented in [2].

The lumped-element circuit, shown in Fig. 3.3, tries to model the inductor through the branch consisting of L_s, R_s, and C_p. The series resistance, R_s, arises from metal resistivity of the spiral inductor and is closely related to the quality factor, being a key issue for inductor modeling. The series feedforward capacitance, C_p, corresponds to the overlap capacitance between the spirals and the underpass metal lines, also called C_l. In order to increase the accuracy of the model, not only the metal spiral but also the *surrounding* environment has to be taken into account. Therefore, both the oxide layer and the silicon substrate have to be modeled. In order to model the oxide layer, a capacitance C_{ox} is set between the spiral and the substrate. Afterwards, in order to model the silicon substrate, a capacitance C_{sub} and a resistance R_{sub} are used. Figure 3.4 illustrates these lumped elements and its relation to the inductor physical implementation.

Fig. 3.3 Typical integrated inductor physical model (π-model)

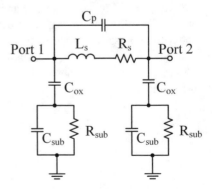

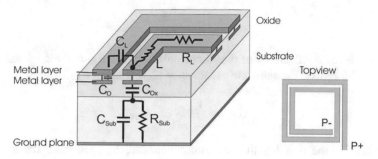

Fig. 3.4 Physical definition of the lumped elements of the inductor model

3.2.1.1 Physical π-Model

As previously mentioned, each lumped element of the π-model can be calculated through an analytical equation. These will be presented in this sub-section.

The resistance R_s, which emulates the series resistance of the inductor, is given by

$$R_s = \frac{\rho \cdot l}{w \cdot \delta \cdot (1 - e^{-t/\delta})} \tag{3.5}$$

where l is the conductor length, ρ is the resistivity of the material, w is the metal width, and t the metal thickness. The skin depth, δ, is given by the following equation [3]:

$$\delta = \sqrt{\frac{\rho}{\pi f \mu}} \tag{3.6}$$

where f is the frequency and μ is the permeability of the metal. The oxide capacitance C_{ox} can be approximated by

$$C_{ox} = \frac{1}{2} \cdot l \cdot w \cdot \frac{\varepsilon_{ox}}{t_{ox}} \tag{3.7}$$

where ε_{ox} is the relative permittivity of the oxide and t_{ox} the oxide thickness.

The silicon substrate is modeled with the resistance R_{sub} and the capacitance C_{sub}. The resistance is used in order to model the resistivity of the substrate, and is given by

$$R_{sub} = \frac{2}{l \cdot w \cdot G_{Sub}} \tag{3.8}$$

where G_{sub} is the conductance per unit area for the silicon substrate and can be approximated by

$$G_{sub} = \frac{\sigma_{S_i}}{h_{S_i}} \qquad (3.9)$$

where σ_{S_i} is the conductivity of the silicon substrate and h_{S_i} is the thickness of the substrate. Moreover, the substrate capacitance C_{sub} is given by

$$C_{sub} = \frac{1}{2} \cdot l \cdot w \cdot C_{ms} \qquad (3.10)$$

where C_{ms} is the unit length capacitance between the metal and the substrate. Finally, the capacitance C_p is calculated through

$$C_p = n_{cross} \cdot w^2 \cdot \frac{\varepsilon_{ox}}{t_{M1-M2}} \qquad (3.11)$$

where t_{M1-M2} is the oxide thickness between the spiral and the underpass metal layer and n_{cross} is the number of crossovers between the spirals and the underpass metal.

The inductance L_s, shown in Fig. 3.3, represents the series inductance and can be calculated through several formulas and techniques [4]. However, one of the most used is the Greenhouse formula [5], given by the following equation:

$$L = 2l \left\{ ln \left[\frac{2l}{w+t} \right] + 0.50049 + \frac{w+t}{3l} \right\} (nH) \qquad (3.12)$$

where L is the series inductance. By using Eqs. (3.5)–(3.12), the designer is able to relate the inductor geometric parameters to the values of the electrical components of the lumped-element circuit given in Fig. 3.3, and, therefore, model, design, and simulate a given integrated inductor with any set of geometric parameters.

3.2.1.2 Physical Segmented Model

However, with the continuing shrinking size of CMOS and with higher operating frequencies, the analytical equations previously presented and the π-model itself are usually too simple to capture all the inductor physics and, therefore, more accurate models have been proposed. The physical model used in this chapter is based on the segmented model approach [6], where the inductor is divided into segments and each segment is characterized with an individual π-model equivalent circuit, as shown in Fig. 3.5.

In this segmented model, the capacitances are calculated by using a more complex method, which is the distributed capacitance model (DCM) [8]. This method provides more accurate results than the analytical equations used in the simple π-model. This DCM method includes high-frequency effects, such as the skin and proximity effects, which, therefore, make the model suitable to be used up to higher frequencies. To calculate the capacitances, we first define the lengths of

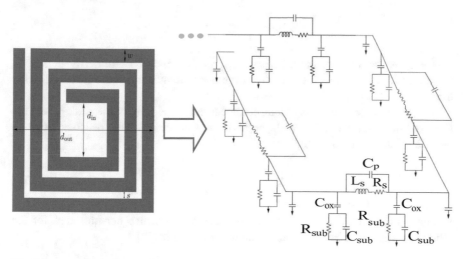

Fig. 3.5 Square inductor, its geometrical parameters and the π-model equivalent circuit for each turn of the inductor (reprinted with permission from [7])

each inductor segment as $l_1, l_2, \ldots, l_n$, and the total length as $l_{tot} = l_1 + l_2 + \cdots + l_n$, where l_n is the length of the last segment of the outer turn. Afterwards, we define,

$$h_k = \sum_{k=1}^{n} \frac{l_k}{l_{tot}} \tag{3.13}$$

and the capacitances can be calculated with the following formulas [8]:

$$C_p = \sum_{k=1}^{n} x \cdot \frac{4}{3} C_{mm} l_k \cdot \frac{\left[(h_k - h_{k-1})^2 + (h_{k+1} - h_k)^2 + (h_{k+1} - h_k)(h_k - h_{k-1})\right]}{(h_{k+1} - h_{k-1})^2} \tag{3.14}$$

$$C_{ox} = \sum_{k=1}^{n} x \cdot \frac{1}{2} \cdot \frac{4}{3} C_{ms} l_k \cdot \frac{\left[(1 - h_{k-1})^2 + (1 - h_k)^2 + (1 - h_k)(1 - h_{k-1})\right]}{3(2 - h_k - h_{k-1})^2} \tag{3.15}$$

An empirical scale factor x, which is the same for both equations, can be used as a fitting parameter in order to adjust the capacitance values to the adopted technology. The 1/2 factor in C_{ox} is due to the fact that the capacitance is divided into two in the π-model. C_{mm} and C_{ms} are the unit length capacitance between the metal spirals and the unit length capacitance between the metal and the substrate, respectively. Normally, these are extracted from measured data, but they can be approximated as follows [9]:

$$C_{mm} = \frac{\varepsilon_0 \varepsilon_{SiO_2} \cdot t}{s} \tag{3.16}$$

$$C_{ms} = \frac{\varepsilon_0 \varepsilon_{SiO_2} \cdot w}{h_{SiO_2}} \qquad (3.17)$$

where ε_0 is the vacuum permittivity, ε_{SiO_2} is the relative permittivity of silicon dioxide, and h_{SiO_2} is the distance from the metal to substrate (the silicon dioxide thickness).

The substrate resistance is crucial for accurately modeling of the peak Q and the shape of the Q curve, along with the series resistance R_s. This resistance can be calculated, according to [10], by

$$R_{sub} = \frac{2}{l \cdot w \cdot G_{sub}} \qquad (3.18)$$

where l is the segment length and G_{sub} is the conductance per unit area for the silicon substrate and can be approximated, according to [9], by

$$G_{sub} = \frac{\sigma_{Si}}{h_{Si}} \qquad (3.19)$$

where σ_{Si} is the conductivity of the silicon substrate and h_{Si} is the thickness of the substrate.

The substrate capacitance can normally be approximated using a simple fringing capacitance model as the one given in [10]. However, to extend our model to high frequencies with more accuracy, we use the DCM technique to calculate the substrate capacitance C_{sub},

$$C_{sub} = \sum_{k=1}^{n} x \cdot \frac{1}{2} \cdot \frac{4}{3} C_{ss} l_k \cdot \frac{\left[(1 - h_{k-1})^2 + (1 - h_k)^2 + (1 - h_k)(1 - h_{k-1})\right]}{3(2 - h_k - h_{k-1})^2} \qquad (3.20)$$

Again, x is the same fitting factor used for previous equations, the $1/2$ factor in C_{sub} is due to the fact that the capacitance is divided into two in the π-model and C_{ss} is the length capacitance between the substrate and the ground plane, which can be approximated by the following equation [9]:

$$C_{ss} = \frac{\varepsilon_0 \varepsilon_{Si} \cdot w}{h_{Si}} \qquad (3.21)$$

By using the previous formulas to calculate the values of R_s, C_p, C_{ox}, C_{sub}, and R_{sub} (given in Eqs. (3.14)–(3.20)) a more reliable inductor model can be developed. Furthermore, the Greenhouse formula, presented in Eq. (3.12), used for L_s, only takes into account the series inductance and not the mutual inductances that appear in integrated inductors, degrading therefore the accuracy of the inductance calculation. Therefore, in order to provide a more accurate inductance calculation, the physical model developed in this chapter uses a set of formulas which describe the individual and mutual inductances of any piecewise structure, therefore being able to calculate the series inductance for any inductor topology.

In 1929, Grover derived formulas for inductance calculation between filaments in several different positions [11]. Greenhouse later applied these formulas to calculate the inductance of a square shaped inductor by dividing the inductor into straight-line segments, as illustrated in Fig. 3.5, and evaluating the inductance by adding up the self inductance of the individual segments and mutual inductance between segments [5]. Some authors call this method the "mutual inductance approach" [12].

Here, the evaluation of the series inductance, L_s, for an octagonal inductor will be given as an example in order to illustrate this mutual inductance approach. Further generalization of the formula for an n-side inductor is also addressed. For hexagonal layouts, the angle between segments is 120°, and for octagonal layouts the angle is 135° and so on. The total inductance of an inductor is given by

$$L_s = L_0 + M_{p+} - M_{p-} - M_{lm} \tag{3.22}$$

where L_s is the total series inductance of the inductor, L_0 is the self inductance of each segment, M_{p+} is the mutual inductances of parallel segments, where the current flows in the same direction whereas M_{p-} accounts for mutual inductance of the parallel segments where currents flow in the opposite directions [5]. Furthermore, M_{lm} accounts for all the different types of mutual inductances resulting from non-parallel segments. These mutual inductances should be summed or subtracted depending on the flow of the current.

For the particular case of the inductor in Fig. 3.6 it is possible to calculate the L_s value through:

Fig. 3.6 Twenty-five segment octagonal spiral inductor layout

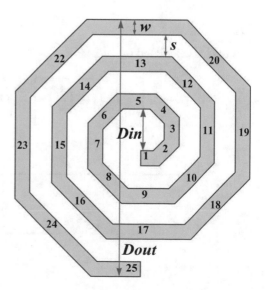

$L_s = L_1 + L_2 + \cdots + L_{25}$ (Self inductance)

$+ 2(M_{1,9} + M_{2,10} + M_{3,11} + M_{4,12} + M_{5,13}$

$+ M_{6,14} + \cdots + M_{17,25})$ (Positive mutual inductances)

$- 2(M_{1,5} + M_{2,6} + M_{3,7} + M_{4,8} + M_{5,9}$

$+ M_{10,6} + \cdots + M_{25,21})$ (Negative mutual inductances)

$- 2(M_{1,2} + M_{2,3} + M_{24,25} + \cdots + M_{1,3} + M_{3,5} + M_{23,25} + \cdots$

$+ M_{1,11} + M_{2,12} + \cdots + M_{16,25})$

(Mutual inductances as shown in Figs. 3.7, 3.8, 3.9, 3.10, and 3.11)

$$(3.23)$$

Due to the magnitude and phase of the currents, the mutual inductances are assumed identical in all sections, hence $M_{a,b} = M_{b,a}$. In order to calculate the self inductance, L_0, the Greenhouse formula given in Eq. (3.12), may be applied. For the evaluation of the mutual inductances, the formulas deducted by Groover [11] are applied. Figure 3.7 represents two parallel segments (e.g., $M_{1,9}$ in Fig. 3.6) by bold lines. Parameter d is the distance between the parallel segments with lengths l and m, and r and q represent the difference between the length of the segments.

For such case, the mutual inductances can be calculated by

$$2M = (M_{m+r} + M_{m+q}) - (M_r + M_q) \qquad (3.24)$$

Each $M_{i,j}$ is calculated with the following formula:

$$M = 2 \cdot l \cdot U \qquad (3.25)$$

where U is calculated by

$$U = ln\left[\frac{l}{d} + \sqrt{1 + \left(\frac{l}{d}\right)^2} \right] - \sqrt{1 + \left(\frac{d}{l}\right)^2} + \frac{d}{l} \qquad (3.26)$$

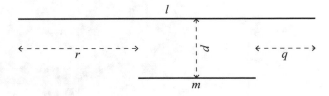

Fig. 3.7 Parallel segments

where d is the distance between segments, which is considered as the geometric mean distance (GMD) between segments and calculated by,

$$\ln(GMD) = \ln(p) - \frac{w^2}{12p^2} - \frac{w^4}{60p^4} - \frac{w^6}{168p^6} - \frac{w^8}{360p^8} - \frac{w^{10}}{660p^{10}} - \cdots \quad (3.27)$$

where p is the pitch of the two wires and w is the width of the segments in study. Note that, for the particular case of a square inductor, the series inductance calculation will not comprise the mutual inductance between two consecutive segments, since their mutual inductance is zero (because the angle between them is 90°). However, for the specific case of the octagonal inductor under study, the angle between them is not 90°, and, therefore, the mutual inductance must be taken into account. An example for mutual inductance for segments which are connected at one end, such as $M_{4,3}$, is presented in Fig. 3.8.

This type of mutual inductance is calculated through [11]:

$$M_{lm} = 2\cos\varepsilon \left[l_1 \tanh^{-1} \left(\frac{m_1}{l_1 + R_1} \right) + m_1 \tanh^{-1} \left(\frac{l_1}{m_1 + R_1} \right) \right] \quad (3.28)$$

In Eq. (3.28), l_1 and m_1 are the lengths of the segments and R_1 is the distance between the segment ends, and can be calculated by

$$R_1^2 = 2l^2(l - \cos\varepsilon) \quad (3.29)$$

It is also possible to use one of the following relations to substitute either R_1 or ε:

$$\cos\varepsilon = \frac{l^2 + m^2 - R_1^2}{2lm} \quad (3.30)$$

$$\frac{R_1^2}{l^2} = 1 + \frac{m^2}{l^2} - 2\frac{m}{l}\cos\varepsilon \quad (3.31)$$

Fig. 3.8 Segments which are connected at one end

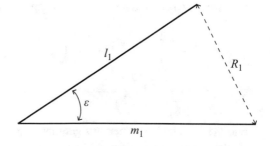

Fig. 3.9 Case for when the
intersection point is lying
outside the two filaments

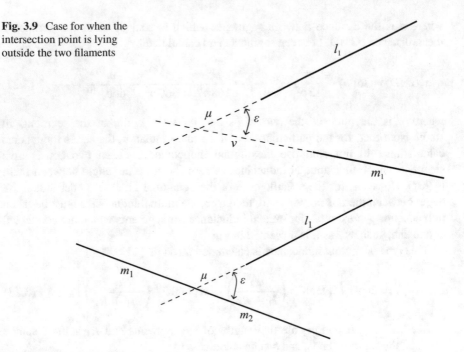

Fig. 3.10 Case for when the intersection point lies upon one filament

The case of mutual inductance where the intersection point is lying outside the two filaments, for example, $M_{3,5}$, is given in Fig. 3.9, whereas the case where the intersection point lies upon one filament, which is the most complex case, such as $M_{5,14}$, is presented in Fig. 3.10.

The mutual inductances in Figs. 3.9 and 3.10 are calculated by the following equation:

$$M_{lm} = 2\cos\varphi \left[(M_{\mu+l,\nu+m} + M_{\mu\nu}) - (M_{\mu+l,\nu} - M_{\nu+m,\mu}) \right] \tag{3.32}$$

The general case for mutual inductances between segments is given in Fig. 3.11, which can be calculated with the following equation:

$$M_{lm} = 2\cos\varepsilon \cdot \left[(\mu + l) \cdot \tanh^{-1}\left(\frac{m}{R_1 + R_2} \right) + (\nu + m) \cdot \tanh^{-1}\left(\frac{l}{R_1 + R_4} \right) \right.$$
$$\left. - \mu \cdot \tanh^{-1}\left(\frac{m}{R_3 + R_4} \right) - \nu \cdot \tanh^{-1}\left(\frac{l}{R_2 + R_3} \right) \right] \tag{3.33}$$

while the so-called intermediary geometric parameters R_1 to R_4 can be calculated with the following set of formulas:

3.2 Modeling Methodologies: Physical vs. Surrogate Models

55

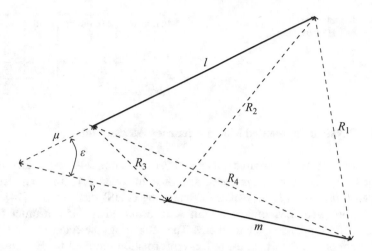

Fig. 3.11 General case for two segments placed in the same plane

$$R_1^2 = (\mu + l)^2 + (\nu + m)^2 - 2(\mu + l)(\nu + m)\cos\varepsilon \tag{3.34}$$

$$R_2^2 = (\mu + l)^2 + \nu^2 - 2\nu(\mu + l)\cos\varepsilon \tag{3.35}$$

$$R_3^2 = (\mu)^2 + \nu^2 - 2\nu\mu\cos\varepsilon \tag{3.36}$$

$$R_4^2 = (\mu)^2 + (m + \nu)^2 - 2\mu(\nu + m)\cos\varepsilon \tag{3.37}$$

The following parameters are required in order to calculate M_{lm} in Eq. (3.33):

$$2\cos\varepsilon = \frac{\alpha^2}{lm} \tag{3.38}$$

$$\alpha^2 = R_4^2 - R_3^2 + R_2^2 - R_1^2 \tag{3.39}$$

$$\mu = \frac{l\left[2m^2(R_2^2 - R_3^2 - l^2) + \alpha^2(R_4^2 - R_3^2 - m^2)\right]}{4l^2m^2 - \alpha^2} \tag{3.40}$$

$$\nu = \frac{m\left[2l^2(R_4^2 - R_3^2 - m^2) + \alpha^2(R_2^2 - R_3^2 - l^2)\right]}{4l^2m^2 - \alpha^2} \tag{3.41}$$

Using Eq. (3.33) and the set of equations presented in (3.34)–(3.41), the inductance of any piecewise structure can be approximated. However it is easy to understand that when one increases the number of segments, the difficulty of calculation increases exponentially due to the number of new mutual inductances that appear [13].

Table 3.1 Inductor variable
ranges for the LHS sampling

Parameter	Minimum	Grid	Maximum
N	1	1	8
D_{in} (μm)	10	1	300
w (μm)	5	0.05	25
s (μm)	2.5	–	2.5

3.2.1.3 Physical Segmented Model Accuracy Assessment

In order to evaluate the accuracy of this physical model, 240 test inductors were generated covering the design space specified in Table 3.1. The samples were generated following a Latin hypercube sampling (LHS) distribution [14]. These 240 inductors were EM simulated with ADS *Momentum* [15] in order to have a benchmark comparison for accuracy. The technology selected was a 0.35 μm CMOS technology, for which the process information required for EM simulation was available. It should be said, that the space between turns, s, was maintained as the minimum size permitted by the technology, because increasing this value does not bring advantages in the inductor performances [16].

The 240 generated inductors were separated by number of turns (30 per turn). The inductors were simulated at two different frequencies with the physical model presented in the previous sections and the performances obtained were compared with the values obtained through EM simulation. An error assessment was performed using the following equation:

$$Error = \frac{EM_{sim} - Model}{EM_{sim}} \times 100 \quad \text{(in \%)} \tag{3.42}$$

It is possible to observe in Table 3.2 that the model errors are very high both for inductance (ΔL) and quality factor (ΔQ), and that the error increases for inductor with higher number of turns. This fact can be explained, by two different facts: on the one hand, inductors with higher number of turns have more segments, increasing the calculation complexity and, therefore, introducing more errors. On the other hand, inductors with higher number of turns are more difficult to model because its behavior changes more abruptly. Therefore, it is possible to conclude that this physical model is not sufficiently accurate for modeling integrated inductors.

The performances of integrated inductors are highly affected by phenomena such as Eddy currents and proximity effects. Even though the segmented model, uses the DCM capacitance model, which already takes into account the Eddy currents and proximity effects, the prediction of these effects is not entirely accurate. Therefore, in some areas of the design space where these effects are dominant, e.g., smaller D_{in} and large w, the error of the model is large. Since the model does not accurately estimate the inductor performances in these areas and in order to increase the accuracy, the design space was reduced to inductors with $D_{in} > 70\,\mu$m and $w < 15\,\mu$m.

Table 3.2 Average error (in %) of inductance and quality factor for 240 test inductors

N	20 MHz		2.5 GHz	
	ΔL (%)	ΔQ (%)	ΔL (%)	ΔQ (%)
1	18.35	61.20	21.55	31.54
2	10.82	43.96	20.44	34.26
3	13.96	40.53	51.09	46.36
4	13.31	35.41	66.40	69.79
5	10.23	25.07	98.64	127.55
6	10.15	24.36	107.93	141.40
7	12.61	27.95	150.34	229.04
8	12.80	28.72	190.42	511.66

Table 3.3 Average error (in %) of inductance and quality factor for 106 test inductors: reduced design space and using fitting factors

N	20 MHz		2.5 GHz	
	ΔL (%)	ΔQ (%)	ΔL (%)	ΔQ (%)
1	3.70	19.62	3.98	5.28
2	1.77	11.75	0.89	3.83
3	1.95	13.26	3.68	9.89
4	2.27	13.28	9.44	11.26
5	2.19	7.62	18.27	15.17
6	2.77	8.77	48.76	87.15
7	3.26	9.64	229.83	354.76
8	3.59	10.49	90.87	146.64

Apart from the design space reduction, several fitting factors (over the analytical equations) were tested in order to reduce the model error. By executing some statistical studies in order to fine tune the physical model, it was found that by setting the fitting factor, x, to the C_p, C_{ox}, and C_{sub} capacitances (in Eqs. (3.14), (3.15) and (3.20)) of $\frac{1}{2}$, the model errors can be decreased, as seen in Table 3.3.

For higher number of turns, the model errors increase at high frequencies (2.5 GHz), which may be due to the fact that inductors with higher number of turns have its SRF around or lower than 2.5 GHz, which means that the inductance and quality factor of inductors are changing very sharply with frequency (see Fig. 3.2). This sharp behavior increases the difficulty of the modeling process, increasing, therefore, the relative error of the model. From the errors shown in Table 3.3, it can be concluded that typical physical models are not suitable for an accurate modeling of integrated inductors; therefore, new techniques are required in order to accurately model these components.

3.2.2 Surrogate Modeling of Integrated Inductors

Surrogate modeling is an engineering method used when an outcome of interest of a complex system cannot be easily (or cheaply) measured either by experiments or simulations [14]. An approximate model of the outcome is used instead. This section first describes the basic steps involved in the generation of a surrogate model, and, afterwards, presents the proposed methodology for modeling inductor performances.

Generating a surrogate model usually involves four steps which are briefly described below, indicating the options adopted in this work.

1. **Design of experiments:**
   The objective of surrogate models is to emulate the output response of a given system. Therefore, the model has to learn how the system responds to a given input. So, the first step in generating surrogate models is to select the input samples from which the model is going to learn. These samples should evenly cover the design space, so that it can be accurately modeled. In order to perform this sampling, different techniques are available, from classical Monte-Carlo (MC) to quasi-Monte-Carlo (QMC) or LHS [14]. In this chapter, LHS is used.
2. **Accurate expensive evaluation:**
   Surrogate models learn from accurate but expensive evaluations. In this chapter, these accurate evaluations are EM simulations, which are performed with ADS *Momentum* [15]. Depending on the size of the training set, these simulations could even last for weeks. However, these simulations are only performed once for a given fabrication technology, therefore being useful for several years, as technology nodes do not become obsolete in months. Any modeling technique can later be used in order to build a new model using the same training set.
3. **Model construction:**
   This concerns the core functions used to build a surrogate model. Literature reports approaches based on artificial neural networks (ANN), support vector machines (SVM), parametric macromodels (PM), Gaussian-process or Kriging models, etc [17]. For the surrogate models developed in this book, ordinary Kriging models are used. Different MATLAB toolboxes like SUMO [18] or DACE [19] support this type of models, being DACE the one finally selected. One of the motivations for using Kriging is that it provides an error estimate. The motivations to select DACE were practical reasons, such as: it is freely available, simple to use, and it runs over the widely used software package MATLAB [20].
4. **Model validation:**
   Many different techniques may be used in order to validate the model and assess its accuracy, e.g., cross-validation, bootstrapping, and subsampling [21]. In this chapter, in order to validate the model, a set of points was generated independently of the training samples. These samples will be referred to as test samples and were also generated using LHS.

3.2.2.1   Modeling Strategy

Inductance and quality factor are functions of the frequency. There have been attempts to build frequency-dependent models [22, 23] for integrated inductors.

However, surrogate models suffer from exponential complexity growth with the number of parameters. This exponential complexity growth is also valid if the number of training samples increases. In order to alleviate this problem, problem-specific knowledge can be exploited. In this book, the modeling of inductors is performed in a frequency-independent fashion. Hence, an independent model is created for each frequency point. This allows to increase the accuracy, highly reduce the complexity of the models and also the time to generate them [24].

The initial strategy to build the model was to create a surrogate model valid in the complete design space. The model was created using 800 inductors generated with the LHS technique. Two different models were developed: one for predicting the inductance and another for the quality factor (denoted as L and Q models, for the sake of simplicity). Recall that, since the models are frequency-independent, L and Q models have to be created for each frequency point. In order to compare the surrogate and the physical modeling techniques, the technology selected was the same $0.35\,\mu m$ CMOS technology, and the test samples used were the same used previously. Therefore, these L and Q models were valid for inductors with any given number of turns N, inner diameter D_{in}, and turn width w for the ranges shown in Table 3.1. After testing the global surrogate model, against the same 240 inductors previously generated to test the analytical model, it is possible to observe in Table 3.4, that the mean relative error is much lower when compared with the physical model. However, the error is still unacceptably large at higher frequencies.

By understanding that the number of inductor turns can only take some discrete values, e.g., in the implementation reported in this book, it can only take integer values, it becomes clear that by creating several surrogate models, one for each number of turns (e.g., one model for inductors with two turns, another for inductors with three turns, etc.) the model accuracy can be increased, because the complexity of the modeled design space decreases. The generation of separate surrogate models for each number of turns instead of considering the number of turns as an input parameter of the surrogate model brings several benefits: not only is the accuracy significantly enhanced but also the computational cost is significantly decreased as the computational complexity of the training process grows exponentially with the number of samples. The number of models to create is manageable as the number of turns is typically between 1 and 8. This strategy increases the overall accuracy and efficiency of the model, as shown for the average errors of 240 test inductors in Table 3.5. However, some test inductors still present large L and Q errors, specifically at high frequencies (around 2.5 GHz) for inductors with many turns.

Table 3.4 Average error (in %) of inductance and quality factor for 240 test inductors: global model for all N

20 MHz		1 GHz		2.5 GHz	
ΔL (%)	ΔQ (%)	ΔL (%)	ΔQ (%)	ΔL (%)	ΔQ (%)
0.03	0.28	0.54	0.62	16.64	3.61

Table 3.5 Inductance and quality factor average error for 240 test inductors (in %): one model for each N

	20 MHz		1 GHz		2.5 GHz	
N	ΔL (%)	ΔQ (%)	ΔL (%)	ΔQ (%)	ΔL (%)	ΔQ (%)
1	0.08	0.46	0.09	0.62	0.08	0.54
2	0.06	0.77	0.14	0.89	0.19	0.78
3	0.03	0.37	0.10	0.52	0.15	0.39
4	0.02	0.32	0.16	0.77	0.24	0.92
5	0.01	0.21	0.07	0.36	0.47	0.61
6	0.01	0.17	0.11	0.50	2.39	1.60
7	0.02	0.34	0.11	0.60	21.70	2.20
8	0.01	0.15	0.11	0.44	6.96	2.36

This high error can be explained by the fact that some inductors from the training set have their SRF below or around the 2.5 GHz range. Kriging surrogate models assume continuity: if an input variable changes by a small amount, the output varies smoothly. However, in the adopted technology, some inductors with more than five turns have their SRF close to 2.5 GHz, where the inductance does not change smoothly (see Fig. 3.2). Therefore, the use of these inductors during the model construction dramatically decreases the accuracy of the model, because they present a sharp behavior and blur the model creation. Therefore, the accuracy estimation of L and Q of these useful inductors is dramatically increased if only inductors with SRF sufficiently above the desired operating frequency are used for model training. However, this option is only feasible if we can detect which inductors have their SRF sufficiently above the frequency of operation and are, hence, useful.

Therefore, in the proposed strategy, the construction of the model is based on a two-step method:

1. Generate surrogate models for the SRF (for each number of turns) using all training inductors.
2. In order to generate highly accurate surrogate models for L and Q, only those inductors from the training set whose SRF is sufficiently above the operating frequency are used. For example, if the operating frequency is 2.5 GHz, only inductors with SRF >3 GHz are used to generate L and Q models. In Fig. 3.2, it is possible to observe that the selected inductors with eight turns are not useful at 2.5 GHz, since their SRF <3 GHz.

Consequently, with this methodology, whenever a test inductor is going to be evaluated, its SRF value is predicted first. If the predicted SRF is below 3 GHz, the inductor is discarded since it is not useful for the selected operating frequency. Otherwise, its inductance and quality factor are calculated using the L/Q models. The algorithm for model training is, therefore, as follows:

Training Phase

Step 1: Sampling
Generate set ITS of B inductor training samples for each number of turns using LHS. Generate set IVS of B_x validation samples for each number of turns using LHS.

Step 2: Simulation
Perform EM simulation of the $(B + B_x)$ samples.

Step 3: SRF Modeling
For each number of turns $i = 1, \ldots, N_{max}$, generate model of self-resonance frequency using set ITS.

Step 4: Inductor Selection
Extract from ITS the set of training inductors ITS^* with $SRF > f_0 + \Delta f$, where f_0 is the frequency of operation and Δf is a safety margin.

Step 5: L/Q Modeling
For each number of turns, $i = 1, \ldots, N_{max}$, generate L/Q model at frequencies of interest using set ITS^*.

Step 6: Validation
Use set IVS to validate the accuracy of the generated models.

Model training for a given technology and frequency of operation is performed only once. This model can be used as many times as needed according to the following algorithm:

Evaluation Phase

For a set IES of C inductors to evaluate $i = 1, \ldots, C$ do:

Step 1: SRF Evaluation
Predict the self-resonance frequency of the i-th inductor using the SRF model.

Step 2
If the predicted self-resonance frequency $SRF_{pre} > f_0 + \Delta f$ go to Step 3. Otherwise, discard i-th inductor as a non-valid inductor for the desired frequency of operation and go to Step 1.

Step 3: L/Q Evaluation
Predict L and Q of i-th inductor at frequencies of interest using the L/Q model.

In order to evaluate the validity of the proposed strategy, the 240 test inductors were evaluated for three different frequencies. The model errors for L and Q are shown in Table 3.6 and for the SRF in Table 3.7. It can be concluded that by following this modeling strategy, the model error for inductance and quality factor is always below 1% for L and Q (even at 2.5 GHz).

Table 3.6 Inductance and quality factor average error for 240 test inductors (in %): one model for each N filtered by SRF

	20 MHz		1 GHz		2.5 GHz	
N	ΔL (%)	ΔQ (%)	ΔL (%)	ΔQ (%)	ΔL (%)	ΔQ (%)
1	0.08	0.46	0.08	0.62	0.08	0.54
2	0.06	0.77	0.14	0.89	0.19	0.78
3	0.03	0.37	0.10	0.52	0.15	0.39
4	0.02	0.32	0.16	0.76	0.24	0.92
5	0.01	0.20	0.07	0.36	0.43	0.54
6	0.02	0.27	0.11	0.49	0.69	0.45
7	0.02	0.69	0.11	0.59	0.93	0.64
8	0.02	0.29	0.12	0.45	0.50	0.98

Table 3.7 SRF average error for 240 test inductors (in %)

	$N = 1$	$N = 2$	$N = 3$	$N = 4$	$N = 5$	$N = 6$	$N = 7$	$N = 8$
SRF	1.16	1.73	0.83	0.64	0.32	0.38	0.42	0.30

3.3 Integrated Inductor Synthesis Methodologies

In this section, several optimization algorithms are used in order to synthesize integrated inductors. The inductor optimization problem can be posed as a constrained optimization problem as described in Eq. (1.1). The synthesis of integrated inductors can be considered as a single-objective problem or a multi-objective problem. In this section, the particle swarm optimization (PSO) algorithm [25] is used for single-objective optimization, and the non-dominated sorting genetic algorithm (NSGA-II) is used for multi-objective optimizations [26].

PSO is a single-objective stochastic optimization technique, which was inspired by nature social behavior. The standard flowchart of PSO is shown in Fig. 3.12.

The standard PSO algorithm is initialized with a set (called swarm) of candidate solutions (called particles). The particles that constitute a swarm move around the search space, with a given *velocity* (which is a parameter used to move from a position to another in the search space) looking for the best solution. Each particle has the ability of saving its current position, its historical best position, and the best position of the neighboring particles [25]. During each iteration of the algorithm, the entire swarm is evaluated by an objective function (for inductors it could be the physical or surrogate model) to determine its *fitness* (the fitness is the value of the objective under optimization). Afterwards, the individual and global bests are updated. At this point, if the maximum number of iterations is achieved, the algorithm stops, if not, the velocity and position of each particle are updated. Each particle adjusts its traveling speed dynamically by taking into account its inertia, its historical best position and that of its best neighbor. These steps are repeated until some stopping condition is met (in this book the stopping condition is always the number of iterations imposed by the user).

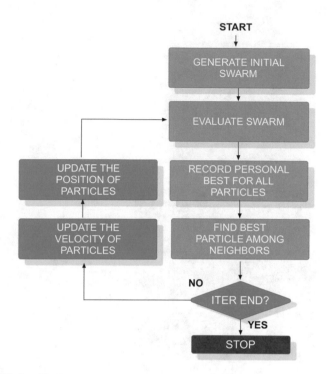

Fig. 3.12 PSO flow chart

The standard PSO algorithm was designed to only deal with unconstrained optimization problems. For inductor synthesis (or any circuit design problem) constraints are a must. Therefore, a tournament selection method [27] has been implemented in PSO to handle design constraints:

1. If two solutions do not comply with constraints, the one with the smallest constraint violation is selected.
2. If one solution complies with constraints (denominated as a feasible point) and another does not (denominated as an infeasible point), the feasible point is selected.
3. If two feasible solutions are compared, the one with the best objective function value is selected.

For multi-objective optimizations, in this book, NSGA-II is adopted [26], whose flow is illustrated in Fig. 3.13.

In the first step an initial population of N individuals (candidate solutions) is generated. Afterwards, these N individuals are evaluated in order to obtain the values of their objectives and constraints. In the next step (assign ranking and crowding distance in Fig. 3.13), a classification process is performed in order to establish which are the most promising individuals. This process is denominated as *rank assignment*. In this process, all individuals are classified into fronts as

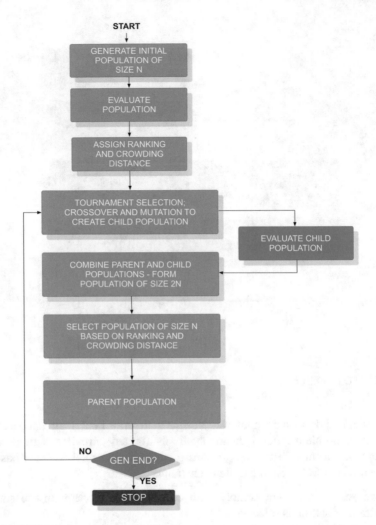

Fig. 3.13 NSGA-II flow chart

follows: all non-dominated individuals are placed in front 1 (F_1); the non-dominated individuals after the individuals in front F_1 are eliminated from the population, and are placed in front 2 (F_2); the non-dominated individuals after the individuals in front F_2 are eliminated from the population, and are placed in front 3 (F_3); and so on. While assigning the ranking, if any individual does not comply with constraints, the constraint violation is used to assign the rank. Afterwards, the crowding distance is also calculated. The crowding distance is a measure of how well the solutions are distributed in the performance space. This measure is especially interesting in multi-objective optimizations because the user is usually looking for a POF as wide and uniform as possible in order to cover all the performance space with different trade-offs.

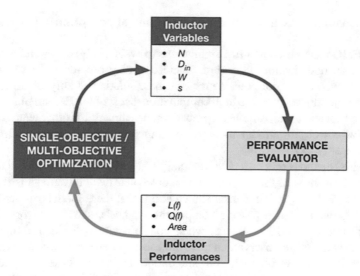

Fig. 3.14 Optimization-based inductor synthesis loop

After assigning ranks and calculating the crowding distance, all the individuals are taken four by four from the population (the so-called parent population), and a tournament selection is performed. This tournament selection consists on a dominance check between pairs of individuals. If the individuals under comparison do not dominate each other, the one with larger crowding distance is selected. The selected individuals are then used to generate new individuals using the crossover operator. Finally, the mutation operator performs random variations on the individuals. The population obtained after selection, crossover, and mutation is denominated child population. This child population has to be evaluated in order to obtain its objectives and constraints, and, after that, it is combined with the previous parent population, in order to form a population with $2N$ individuals. After that, a survivor selection is performed by assigning the rank and calculating the crowding distance for the individuals. A new parent population is obtained by selecting individuals with higher rank and larger crowding distance. These steps are repeated until the number of generations imposed by the user is reached.

The synthesis of integrated inductors shown here does not exploit any specific characteristics of PSO and NSGA-II, hence, they can be replaced by any other single-objective and multi-objective optimization algorithms, respectively. The basic optimization-based synthesis flow for integrated inductors is illustrated in Fig. 3.14. The synthesis strategies can be classified into four categories according to the kind of performance evaluator and how it interacts with the optimization technique. They will now be discussed in detail, highlighting their advantages and drawbacks:

1. **EM simulation as a performance evaluator of an optimization technique (EMO):**
   The EMO methods embed EM simulations to evaluate objectives and constraints within an optimization algorithm. Therefore, they provide the most accurate performance evaluation and, hence, the best solution quality of all methods. Their major drawback is the high computational cost of the EM simulations. This method constitutes an excellent comparison benchmark for other techniques.

2. **Equivalent circuit model as a performance evaluator of an optimization technique (ECO):**
   The ECO methods rely on a physical/analytical equivalent circuit model to obtain the performances of the passive component. Their main advantage is their high efficiency. However, it has been shown above that these models are usually not accurate enough for a passive component synthesis, showing errors typically much higher than 10%. Hence, when coupled with optimization algorithms, large deviations are observed on the synthesized passive components when their performances are verified with EM simulation. Therefore, these models can only be used for a first order approximation and not for a full inductor synthesis.

3. **Offline surrogate model as a performance evaluator of an optimization technique (OFFSO):**
   With the OFFSO methodology, a global surrogate model is created before being used within an optimization algorithm [28]. The surrogate model is first built to be as accurate as possible. Then, the optimization algorithm uses this surrogate model as the performance evaluator to find the optimal solution. The surrogate model is called offline because training data can be generated by EM simulation and the model can be constructed before any optimization objectives and/or constraints are set. Normally, the data set used to build the surrogate model is generated covering the entire design space. When combined with an optimization algorithm, this method has the ability of searching through the entire design space in order to find a global optimum. The generation of the training data is computationally expensive. However, such training data is generated only once and is valid for any future optimization problem. Moreover, they can be generated offline, much before they are needed for the first inductor optimization problem. On the other hand, since the model can be evaluated very fast, the optimization process itself is highly efficient, usually in the range of few minutes.

4. **Online surrogate model as a performance evaluator of an optimization technique (ONSO):**
   Since global surrogate models may be locally inaccurate, ONSO methodologies, also known as surrogate-assisted evolutionary algorithms (SAEA), have received considerable attention, also in practical analog/RF circuit design problems [29]. In this methodology, a coarse surrogate model using a few training points is first constructed. Then, this coarse model is coupled with an optimization algorithm and promising solutions (typically one) are electromagnetically simulated at each iteration of the optimization loop. The data from this EM simulation is used to update the surrogate model to make it more accurate in the region where new simulation points are added, while moving towards the presumed optimal

inductor. However, the outputs of the ONSO methodology highly depend on the accuracy of the initial coarse model, which leads to two significant challenges for this methodology. First, the promising solutions found at the different iterations define the search space and the constructed surrogate model is only accurate in that space. Second, the success of ONSO comes from the basic assumption that the optimal point of the coarse and fine models is not far away in the design space. However, again, this assumption only holds when the coarse model is accurate enough. ONSO methods exhibit a delicate trade-off between efficiency and the probability of convergence to the global optimum. Better convergence can be achieved by either increasing the amount of training data of the coarse model, or emphasizing the exploration of potentially good regions of the design space during the optimization process, or a combination of both. In all cases, an increase in the number of EM simulations is implied, diluting in this way the efficiency advantage over EMO methods. A prescreening technique that can be used in ONSO methods in order to increase accuracy, consists in using the uncertainty measurement of the prediction, i.e., the mean square error (MSE), instead of just the predicted value to rank promising solutions. Such techniques include methods like lower confidence bound, probability of improvement, or expected improvement (EI). The EI method, which is going to be used later in this chapter, uses the MSE in order to evaluate areas of the design space that could be optimal considering also the model error. Therefore, new promising solutions can be found by taking into account the MSE of the model.

In the following sections, these approaches are examined and compared for the synthesis of integrated inductors.

3.3.1 Experimental Results: Single-Objective Optimizations

In this section, four different methodologies: EMO, OFFSO, ONSO, and ONSO using expected improvement (for the sake of simplicity, further references to this method are denoted as ONSOEI) are applied to the synthesis of integrated inductors. The ECO methodology is not implemented since it was shown in Sect. 3.2.1.3 that physical models are not accurate at all, and, therefore, the optimization process yields suboptimal, and even invalid, inductors. The EMO method uses ADS *Momentum* as a performance evaluator and the OFFSO methodology uses the previously developed surrogate model (using the independent N modeling and the SRF filtering strategy). The ONSO and ONSOEI methodologies have some differences when compared to the previous methods. The flow diagram of ONSO/ONSOEI is presented in Fig. 3.15.

An initial surrogate coarse model is created using 40 (EM simulated) training inductors. In this implementation of ONSO, the model is updated at each iteration with the EM simulation results of the best individual of the current population. In ONSOEI, the model is updated by simulating the inductor that presents the

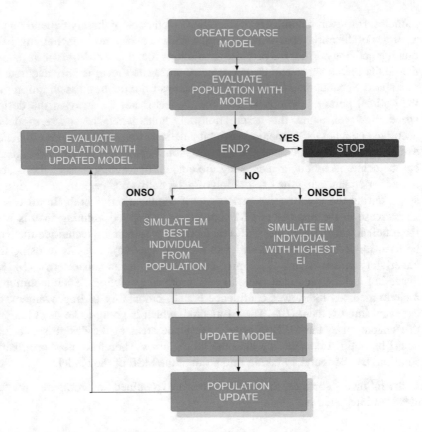

Fig. 3.15 Flow diagram of ONSO/ONSOEI

highest expected improvement from the current population. However, if any of these individuals (either the best one from the current population or that with the highest expected improvement) had already been simulated and used for the model construction in previous iterations, this individual is not EM simulated and the model does not have to be updated in that iteration. The technology used for inductor synthesis was the same 0.35 μm CMOS technology used for the model comparisons in Sect. 3.2. The bounds of the optimization search space are the same bounds of the samples used to create the surrogate models, and shown in Table 3.1.

The optimization constraints were defined to guarantee the "good behavior" of the inductor at the frequencies of interest. The inductors have to be designed in such a way that they are in the so-called flat-bandwidth (BW) zone, as shown in Fig. 3.16 [16]. In order for the inductor to be in this zone, the inductance value must be sufficiently flat from DC (L_{DC}) to slightly above the operating frequency ($L_{eq}(f_o + \Delta f)$, where f_o is the operating frequency) and the self-resonance frequency must lie sufficiently above this frequency. Since in the EMO methodology it is extremely expensive to accurately calculate the SRF of the inductors, the latter

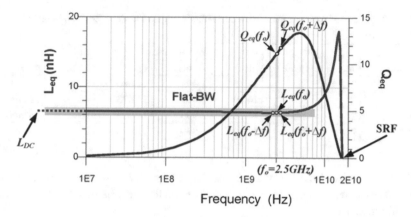

Fig. 3.16 Flat-bandwidth zone (Reprinted with permission from [16])

constraint is approximated by imposing that the quality factor at the operating frequency ($Q_{eq}(f_o)$) is near its maximum and always with a positive slope around it (($Q_{eq}(f_o + \Delta f) - Q_{eq}(f_o)$) > 0) [16]. In order to constrict the area, a constraint is added so that the inductor fits within a 400 μm × 400 μm square. Therefore, the optimization constraints are formulated as

$$area < 400\,\mu m \times 400\,\mu m \tag{3.43}$$

$$\left| \frac{L_{at2.5\,GHz} - L_{at2.55\,GHz}}{L_{at2.5\,GHz}} \right| < 0.01 \tag{3.44}$$

$$\left| \frac{L_{at2.5\,GHz} - L_{at2.45\,GHz}}{L_{at2.5\,GHz}} \right| < 0.01 \tag{3.45}$$

$$\left| \frac{L_{at2.5\,GHz} - L_{at0.1\,GHz}}{L_{at2.5\,GHz}} \right| < 0.05 \tag{3.46}$$

$$Q_{at2.55\,GHz} - Q_{at2.5\,GHz} > 0 \tag{3.47}$$

Equation (3.46) is used in order to guarantee that the inductor is in the flat-BW zone, by ensuring less than 5% deviation between 0.1 GHz and f_o (in this example set at 2.5 GHz). Eqs. (3.44) and (3.45) guarantee that the inductance is very flat around f_o.

Furthermore, Eq. (3.47) is used in order to guarantee a positive slope in the quality factor around f_o, which is used to ensure that the SRF is still far away. To have a fair comparison with the other methods, and although it was not strictly necessary (because an SRF model was developed with the surrogate), the constraint related to the SRF location was imposed identically in all methods. The objective of the first test is to find an octagonal inductor with $L_{spec} = 2nH$ at 2.5 GHz while maximizing the quality factor. Since PSO is a single-objective algorithm, a

Table 3.8 Results of one execution with all methods targeted at $L = 2nH$ and maximizing Q. Area $<400\,\mu m \times 400\,\mu m$

Method	N	D_{in} (μm)	w (μm)	Performances L (nH)	Q	EM sims.	CPU time
EMO	2	225	14.15	1.999	11.086	8000	106.5 h
ONSO	2	225	14.10	1.999	11.073	45	32 min
ONSOEI	2	225	14.10	1.999	11.073	55	56 min
OFFSO	2	225	14.10	1.999	11.073	0	2 min

weighted objective function was built so that the quality factor was maximized and the difference between the desired inductance and the one obtained by the algorithm was minimized:

$$f(x) = -Q(x) + \lambda |L(x) - L_{spec}| \qquad (3.48)$$

where λ is set to 5. Results for the best inductor obtained by a single execution of each methodology are shown in Table 3.8. The performances of such optimal inductors have been electromagnetically simulated a posteriori, so that the L and Q values shown at the table can be fairly compared.

It must also be considered that the time data in Table 3.8 correspond to CPU time. The computer used for all simulations had two six-core Intel® Xeon® E5-2630 v2 processors at 2.60 GHz, enabling parallelization of the evaluation of different solutions. Such parallelization is very effective for the EMO approach, achieving a reduction of elapsed time by about a 10× factor with respect to the CPU time. A priori, parallelization of the OFFSO approach is not worth given the total CPU times involved. Model updating in ONSO and ONSOEI is performed with at most one inductor at each iteration, and, therefore such parallelization is not directly possible.

Notice that the number of EM simulations, and, therefore, the computation times for the OFFSO, ONSO, and ONSOEI methods only include the simulations performed during the execution of the PSO algorithm. Construction of the initial coarse model in ONSO and ONSOEI requires another 40 simulations (5 for each number of turns). This takes another 23.11 h of CPU time. The CPU time required to simulate the 800 training samples in OFFSO amounts to 462.30 h (around 19 days). Indeed, as stated above, parallelization of these evaluations in the machine with two six-core processors reduces the elapsed time by approximately a factor 10x with respect to the CPU time. Since the initial 40 EM simulations for the coarse model in ONSO and ONSOEI and the 800 samples for the OFFSO method can be run a priori and it is a one-time investment (they are independent of the required optimization objectives), they have not been included in the CPU time calculation of the optimization process shown in the table. The table only reflects the CPU time required to get the optimization results once the optimization goals are known. This means that the initial accurate simulations should not be accounted as time to build a given model as they are only performed once and can be used to build several different models using different techniques. To understand the lack of proportionality between the CPU times above and those in the table, two considerations must be made:

1. Some simulations for the initial sampling take longer, since simulations of inductors with eight turns require much more time than, e.g., three turns, whereas convergence of the different methods in this example implies that most additional EM simulations correspond to inductors with less than four turns
2. The EM simulations of the 40 samples of the coarse model in ONSO and ONSOEI and the 800 samples in OFFSO were performed for several hundreds of frequency points so that the results can be used for any synthesis problem independently of the frequencies of interest. If only the few frequency points necessary for this experiment were used, the CPU time is reduced from 23.11 to 1.22 h, and from 462.30 to 24.36 h, respectively. However, in that case, new samples and their corresponding EM simulation are needed whenever a new frequency of operation is desired.

As an illustration example, the performances of the inductor obtained with the OFFSO methodology in the first test example (Table 3.8) are presented in Fig. 3.17 and they are compared against the EM simulation of the same inductor. It is possible to observe that the accuracy of the model is remarkably good along all the frequency range. After the optimization process (which takes 2 min) the inductor synthesis is complete and the layout of this inductor is presented in Fig. 3.18.

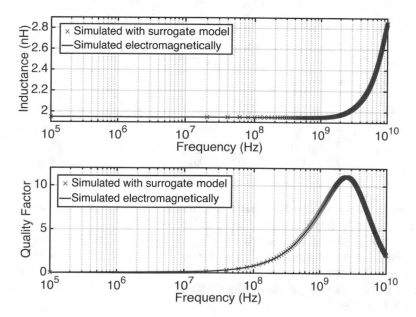

Fig. 3.17 Performance parameters of the 2nH inductor obtained in the first test example: inductance and quality factor vs frequency curves. (Reprinted with permission [24])

Fig. 3.18 Layout of the 2nH inductor obtained in the first test example. (Reprinted with permission [24])

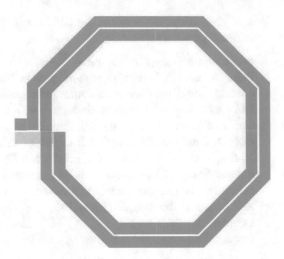

Table 3.9 Results of one execution with all methods targeted at $L = 2.5nH$ and maximizing Q. Area $<400\,\mu m \times 400\,\mu m$

Method	N	D_{in} (μm)	w (μm)	Performances L (nH)	Q	EM sims.	CPU time
EMO	2	260	10.60	2.504	10.704	8000	111.2 h
ONSO	3	136	9.50	2.500	10.390	36	34.3 min
ONSOEI	2	259	10.35	2.499	10.713	59	67.6 min
OFFSO	2	260	10.80	2.498	10.742	0	2 min

Table 3.10 Mean values and standard deviations (between brackets) with all methods targeted at $L = 2nH$ and maximizing Q. Area $<400\,\mu m \times 400\,\mu m$

Method	Performances L (nH)	Q	EM sims.	CPU time
EMO	1.999 (0.000)	11.08 (0.000)	8000 (0.0)	106.5 (0.0) h
ONSO	1.999 (0.000)	11.073 (0.000)	45 (0.0)	32 (0.0) min
ONSOEI	1.999 (0.000)	11.073 (0.000)	55 (0.0)	56 (0.0) min
OFFSO	1.999 (0.000)	11.073 (0.000)	0	2 (0.0) min

The results for another experiment, this time addressing an inductor with 2.5nH and maximum quality factor at 2.5 GHz, are shown in Table 3.9.

As in any other computational intelligence algorithm, PSO also implies the introduction of randomness, and, hence, different runs may provide different results. Therefore, 20 independent runs were performed for each experiment. The mean values and standard deviations obtained with the different techniques are shown in Tables 3.10 and 3.11.

From the comparison of these results of the different techniques for single-objective optimization, several conclusions can be drawn:

Table 3.11 Mean values and standard deviations (between brackets) with all methods targeted at $L = 2.5nH$ and maximizing Q. Area $<400\,\mu m \times 400\,\mu m$

| Method | Performances | | EM | CPU |
	L (nH)	Q	sims.	time
EMO	2.502 (0.001)	10.650 (0.102)	8000 (0.0)	111.2 (0.0) h
ONSO	2.500 (0.001)	10.390 (0.335)	35.9 (8.0)	34.3 (8.3) min
ONSOEI	2.499 (0.008)	10.713 (0.265)	62.2 (33.1)	67.6 (36.0) min
OFFSO	2.498 (0.010)	10.742 (0.144)	0	2 (0.0) min

1. OFFSO converges to approximately the same solutions than the reference method EMO in all cases. However, the efficiency is drastically increased, since the CPU time decreases from more than 100 h to roughly 2 min. These test examples demonstrate that the accuracy of the surrogate model used in the OFFSO methodology leads to the same design space areas with much higher efficiency.
2. All methods practically converge to the same solutions in the 2nH example but a larger variability between different executions appears for the 2.5nH example in ONSOEI and especially ONSO, yielding a slightly lower quality factor. This seems to indicate that despite the EM simulations used in order to iteratively increase the accuracy of ONSO, the accuracy of the initial coarse model is crucial for an optimal inductor synthesis, i.e., if during the first iterations, where the model is still quite inaccurate, the optimization algorithm leads to design space areas away from the optimal point, it is less likely that the optimization algorithm will converge to the global optimum. Moreover, the computation time rises to tens of minutes due to the additional EM simulations. The use of the expected improvement in ONSOEI improves, in most cases, the ONSO ability to locate the optimal regions of the search space and, therefore, the accuracy of the method is quite good. However, the use of this prescreening method brings a penalty in the form of additional EM simulations that typically rises the optimization time to about 1 h.

In the examples shown above, the OFFSO, ONSO, and ONSOEI methodologies converge to the same solutions either if the training inductors with SRF close or below the frequency of operation are filtered out according to the two-step modeling method described above or not. Inductors with a number of turns $N \leq 3$ are obtained. All training inductors with these numbers of turns have a SRF sufficiently above 2.5 GHz. Therefore, it becomes natural that the effect of the SRF filtering methodology previously proposed is negligible. However, when the specs are such that they are met with inductors with a larger number of turns, they tend to have smaller SRF and the effect of the proposed modeling approach becomes more noticeable. To show this, we will consider the comparison to methods similar to OFFSO, ONSO, and ONSOEI, but in which all inductors are used in the training phase of the L/Q model, i.e., no inductor is filtered out if its SRF is close to

Table 3.12 Results of one execution with all methods targeted at $L = 2.9nH$ and maximizing Q. Area <125 μm × 125 μm

Method	N	Din (μm)	w (μm)	Feasible solutions found?	Performances L (nH)	Q	EM sims.	Constraints met after EM sim?	CPU time
EMO[a]	6	36	5.15	Yes	2.90	8.45	8000	Yes	418.5 h
ONSO	6	35	5.45	Yes	2.89	8.49	9	Yes	32.2 min
ONSO$_{NF}$	4	70	5	No	–	–	2	–	4.2 min
ONSOEI	6	35	5.45	Yes	2.89	8.49	45	Yes	111.2 min
ONSOEI$_{NF}$	6	35	5.45	Yes	2.89	8.49	45	Yes	111.2 min
OFFSO	6	37	5.3	Yes	2.98	8.55	0	Yes	2 min
OFFSO$_{NF}$	6	40	5	Yes	3.11	8.59	0	No	2 min

[a]The elapsed time in EMO reduces by a factor 10 with respect to the CPU time when all cores of the twin six-core processor are used

or below 2.5 GHz. Correspondingly, we will denote these methods as OFFSO$_{NF}$, ONSO$_{NF}$, and ONSOEI$_{NF}$.

In a first example, an inductor of 2.9nH with maximum quality factor that fits into a 125 μm × 125 μm square is requested. Identical performance constraints to the previous examples are imposed. The results for one execution are shown in Table 3.12. In this case, OFFSO and OFFSO$_{NF}$ arrive at different results. Moreover, when the resulting inductors are EM simulated, the performance constraints in OFFSO$_{NF}$ are not met any more. ONSO, ONSOEI, and ONSOEI$_{NF}$ arrive at results similar to OFFSO, but ONSO$_{NF}$ is unable to converge to a feasible solution (as indicated in the fifth column in Table 3.12). A second example is shown in Table 3.13. In this case an inductance of 4.6nH with maximum quality factor within a 140 μm × 140 μm square is specified. In this case, all methods but EMO and OFFSO are unable to converge to a solution due to the larger errors of the surrogate models of inductance and quality factor.

As in the previous experiments, 20 executions of each algorithm were performed for both cases. The statistical analysis of the results is shown in Tables 3.14 and 3.15. The second column in both Tables 3.14 and 3.15 shows how many of the 20 executions the optimization algorithm found a feasible solution. The statistical analysis of L and Q has been performed only for the feasible solutions found. It can be checked in Table 3.14 that ONSOEI arrives to similar solutions than OFFSO, although with a penalty in the computation time. ONSO$_{NF}$ never arrives at a feasible solution, like in the single execution shown in Table 3.12. It is also found that ONSO and ONSOEI$_{NF}$ only converge to a solution about half of the executions. Table 3.15 shows that only EMO and the proposed OFFSO approach always converge to a solution and they are quite similar, but with orders of magnitude less computational effort in the OFFSO case.

Table 3.13 Results of one execution with all methods targeted at $L = 4.6nH$ and maximizing Q. Area $<140\,\mu m \times 140\,\mu m$

Method	N	Din (μm)	w (μm)	Feasible solutions found?	Performances			Constraints met after EM sim?	CPU time
					L (nH)	Q	EM sims.		
EMO[a]	7	40	5.05	Yes	4.56	8.616	8000	Yes	590.5 h
ONSO	5	70	5	No	–	–	3	–	6.8 min
ONSO$_{NF}$	5	70	5	No	–	–	3	–	7.5 min
ONSOEI	6	55	5	No	–	–	5	–	12.4 min
ONSOEI$_{NF}$	5	70	5	No	–	–	9	–	11.4 min
OFFSO	7	40	5	Yes	4.54	8.520	0	Yes	2 min
OFFSO$_{NF}$	6	55	5.05	No	–	–	0	–	1.5 min

[a]The elapsed time in EMO reduces by a factor 10 with respect to the CPU time when all cores of the twin six-core processor are used

Table 3.14 Mean values and standard deviations (between brackets) with all methods targeted at $L = 2.9nH$ and maximizing Q. Area $<125\,\mu m \times 125\,\mu m$

Method	Feasible solutions found?	Performances			Constraints met after EM sim?	CPU time
		L (nH)	Q	EM sims.		
EMO[a]	20/20	2.904 (0.000)	8.421 (0.000)	8000 (0.0)	20/20	418.5 (0.0) h
ONSO	11/20	2.891 (0.000)	8.488 (0.000)	9.33 (6.02)	11/20	32.2 (0.0) min
ONSO$_{NF}$	0/20	–	–	2 (0.0)	0/20	4.2 (0.0) min
ONSOEI	20/20	2.890 (0.002)	8.483 (0.022)	45 (0.0)	20/20	111.2 (0.0) min
ONSOEI$_{NF}$	13/20	2.889 (0.004)	8.422 (0.182)	37.1 (15.1)	13/20	95.2 (37.5) min
OFFSO	20/20	2.986 (0.004)	8.545 (0.018)	0	20/20	2 (0.0) min
OFFSO$_{NF}$	20/20	3.110 (0.000)	8.590 (0.000)	0	0/20	1.5 (0.0) min

[a]The elapsed time in EMO reduces by a factor 10 with respect to the CPU time when all cores of the twin six-core processor are used

From the latter experiments, it can be concluded that in many cases the proposed two-step surrogate modeling strategy plays a key role in the proper convergence and accuracy of the surrogate-based optimization techniques for inductor synthesis.

3.3.2   Experimental Results: Multi-Objective Optimizations

One of the advantages of having a fast and accurate surrogate model that needs no EM simulations during the optimization process is the ability of using this model within a multi-objective optimization algorithm. Given the lower maturity of ONSO-like multi-objective optimization techniques and the fact that already for single-objective optimization of inductors, the ONSO and ONSOEI methodologies

Table 3.15 Mean values and standard deviations (between brackets) with all methods targeted at $L = 4.6nH$ and maximizing Q. Area $<140\,\mu m \times 140\,\mu m$

Method	Feasible solutions found?	Performances			Constraints met after EM sim?	CPU time
		L (nH)	Q	EM sims.		
EMO[a]	20/20	4.562 (0.017)	8.564 (0.131)	8000 (0.0)	20/20	590.5 (0.0) h
ONSO	0/20	–	–	4.15 (2.5)	0/20	9.4 (4.4) min
ONSO$_{NF}$	0/20	–	–	3 (0.0)	0/20	7.1 (0.4) min
ONSOEI	0/20	–	–	5 (0.0)	0/20	12.4 (0.0) min
ONSOEI$_{NF}$	0/20	–	–	6.8 (2.0)	0/20	12.0 (0.5) min
OFFSO	20/20	4.540 (0.007)	8.530 (0.029)	0	20/20	2 (0.0) min
OFFSO$_{NF}$	0/20	–	–	0	0/20	1.5 (0.0) min

[a]The elapsed time in EMO reduces by a factor 10 with respect to the CPU time when all cores of the twin six-core processor are used

have not provided better solutions than the OFFSO approach and always with a higher computation time, this section will compare the OFFSO and EMO methodologies with multi-objective optimization algorithms. The effect of the two-step modeling approach will be also studied. The search space is the same used in the previous optimizations and shown in Table 3.1. As a first test example, an optimization with two objectives was performed. The optimization problem had two objectives: inductance and quality factor maximization. Furthermore, the inductors should comply with the constraints described in Eqs. (3.43)–(3.47).

This optimization was performed with 300 individuals and 100 generations. The results are presented in Fig. 3.19, where the results of the optimization using the surrogate model have been electromagnetically simulated so that the accuracy can be fairly compared. It is possible to observe that the Pareto fronts obtained by both methods are very similar. The advantage of using OFFSO is the efficiency. While with EMO the optimization lasted 355.55 h (around 15 days CPU time),[1] the OFFSO method lasted 4 min, which is an increase in efficiency of about three orders of magnitude.

A second test example was an optimization with three objectives, this time with 1000 individuals and 80 generations. Since area is of great importance especially in IC technologies, area minimization was added as a third objective in the optimization. A major motivation for selecting a large population size is that the application of inductor fronts to RF circuit design benefit from denser Pareto fronts [30]. The results are shown in Fig. 3.20. The results of the surrogate model that are shown in Fig. 3.20 have been electromagnetically simulated a posteriori so that both fronts can be fairly compared.

---

[1]The elapsed time reduces to 29 h with respect to the CPU time when all cores of the twin six-core processor are used.

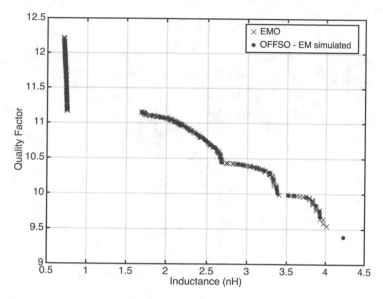

Fig. 3.19 Pareto-optimal front for a two-objective optimization, maximizing quality factor and inductance (Reprinted with permission from [24])

Again, the Pareto fronts achieved by both methods are very similar, and their comparison is not easy. For single-objective optimization algorithms, comparing two solutions given by two different optimization processes is simple because one just has to compare the fitness function (the objective value). For 2D problems, by drawing two given POFs and inspecting them (e.g., Fig. 3.19) it is still relatively easy to see which POF dominates the other. However, this is not the case for 3D multi-objective algorithms, where hundreds of solutions are given as a POF in a 3D space. Therefore, in order to perform a comparison between POFs, a more accurate comparison should be performed by using performance metrics, specifically developed for this purpose. Throughout this book, two different metrics will be used: hypervolume and coverage set. The hypervolume metric gives an insight on diversity and convergence of the POF, while the coverage set allows performing binary comparisons between two fronts, as it gives the percentage of points of each front dominated by the other one [31].

The hypervolume is basically a calculation of the union of the hypercubes determined by each solution in the objective space and a reference point (see Fig. 3.21). A feature of the hypervolume metric is that it does not require any knowledge of the true Pareto front, that is especially convenient in this engineering problem in which the true Pareto front is not known. The hypervolume metric depends on the selected reference point, hence, the same reference point must be used in order to fairly compare the Pareto fronts generated with the two different techniques.

Furthermore, given two solution sets, P_1 and P_2, the set coverage is defined as

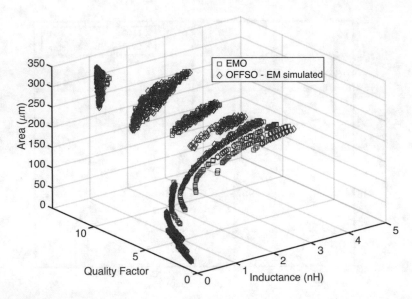

Fig. 3.20 EM simulation of Pareto-optimal fronts for a 3 objective optimization, maximization of quality factor and inductance and minimization of area using EMO and OFFSO methods (Reprinted with permission from [24])

Fig. 3.21 Illustrating the hypervolume calculation

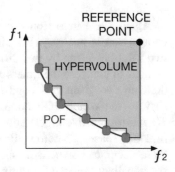

$$C(P_1, P_2) = \frac{|\{b \in P_2 | \exists a \in P_1, a \prec b\}|}{|P_2|} \qquad (3.49)$$

where $a \prec b$ means that a dominates b. A similar value of $C(P_1, P_2)$ and $C(P_2, P_1)$ implies that no front is better than the other. In practice, given two solution sets, P_1 and P_2, the set coverage is defined as $C(P_1, P_2)$, which is the ratio of solutions in P_2 that are dominated by at least one solution in P_1, e.g., if $C(P_1, P_2)=1$, it means that all solutions in P_2 are dominated by P_1.

The hypervolume of the Pareto front obtained with the EMO approach was $HV_{EMO} = 10,674$ and that with the OFFSO approach (after electromagnetically simulating the final results) was $HV_{OFFSO} = 10,604$, which is very similar. Regarding the set coverage, the calculated figures were

$$C(\text{PF}_{\text{EMO}}, \text{PF}_{\text{OFFSO}}) = 0.16$$

$$C(\text{PF}_{\text{OFFSO}}, \text{PF}_{\text{EMO}}) = 0.14$$

that indicates that practically the same percentage of points are dominated by the other front.

The EMO optimization took 1926.39 h (roughly 80 days CPU time)[2] while the OFFSO lasted 7 min, which is an incredible increase in efficiency while obtaining very similar Pareto fronts due to the accuracy of the surrogate model. It is also interesting to compare the results with those of the optimization using the surrogate model without the first filtering stage, i.e., the OFFSO$_{NF}$ methods. Figure 3.22 compares the results of the EMO approach and the OFFSO$_{NF}$ approach (after electromagnetically simulating the final results). There are areas of the Pareto front that are not found with the surrogate-based optimization (infeasible points are not plotted). These areas mostly correspond to inductors with larger number of turns, where the SRF is considerably lower and the SRF filtering approach becomes more noticeably for the 2.5 GHz operating frequency.

The hypervolume of the Pareto front obtained now with the OFFSO$_{NF}$ approach (after electromagnetically simulating the final results) was HV$_{OFFSO_{NF}} = 9626$,

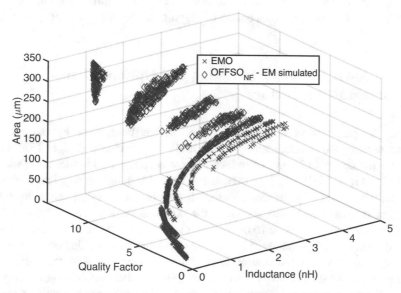

Fig. 3.22 EM simulation of Pareto-optimal front for a 3 objective optimization, maximization of quality factor and inductance and minimization of area using EMO and OFFSO$_{NF}$ methods (Reprinted with permission from [24])

---

[2]The elapsed time reduces to 144 h when all cores of the twin six-core processor are used.

which is clearly inferior to the EMO and OFFSO approaches. Regarding the set coverage, the calculated figures were

$$C(\text{PF}_{\text{EMO}}, \text{PF}_{\text{OFFSO}_{NF}}) = 0.22$$
$$C(\text{PF}_{\text{OFFSO}_{NF}}, \text{PF}_{\text{EMO}}) = 0.13$$

that indicates that the quality of the surrogate model has clearly decreased. It is possible to conclude that the application of surrogate modeling strategies can enhance the efficiency-accuracy trade-off of conventional analytical or EM-based inductor optimization techniques for RF integrated circuits. The two-step surrogate modeling strategy dramatically improves the modeling accuracy of integrated inductors.

It can be concluded that prescreening techniques can be used in surrogate-assisted optimization techniques in order to achieve similar results to EM-based approaches with a significantly lower computation time. However, it is also found that similar or better results are obtained if offline surrogate models with a sufficient number of inductor samples are created following the proposed modeling strategy. Much lower CPU times must be invested during the optimization process since expensive EM simulations are not performed during the optimization stage. The quality of the results of the proposed strategy is also competitive in multi-objective optimization problems, whereas orders of magnitude computation time is saved.

3.4 SIDe-O: A Tool for Modeling and Optimization of Integrated Inductors

During the past few years, an immense effort has been made by the research community for the development of CAD tools for RF circuit design [32]. CAD tools for the design of integrated inductors have been reported in literature, such as ASITIC [33] or SISP [6]. However, these tools are based on physical/analytical models, which typically present accuracy issues in some design space areas and at higher frequencies, as was shown in the previous section. Currently, foundries and EDA companies provide tools for inductor design and optimization, but with some drawbacks: either analytical models are used to model inductors, presenting accuracy issues, or limited optimization options are provided to the user. More commercial tools such as Ansys VeloceRF [34] are also available on the market. This tool provides accurate models based on an electromagnetic simulator engine (more expensive simulations) and the tool allows inductor optimization, but also with limited options (only maximization of the quality factor). Furthermore, no available tool allows multi-objective optimizations, which may be very useful nowadays for obtaining design trade-offs for a given technology and performing design space exploration.

In this section a MATLAB toolbox, called SIDe-O, is presented [35]. This toolbox provides accurate inductor models (based on the surrogate modeling techniques previously presented), diverse and complex optimization options and was the first tool to allow multi-objective optimization of integrated inductors.

3.4.1   Designer Interface

In this section, only an introduction to the designer interface of the toolbox is given, as the fundamental techniques have been described in previous sections. The complete designer interface can be observed in Fig. 3.23. The surrogate models used in the toolbox were developed in MATLAB, therefore, for a straightforward integration, the graphical user interface (GUI) was also developed in MATLAB. The GUI is multi-tabbed, with each tab suited for a different operation.

The first tab in Fig. 3.23a, *Inductor Simulation*, allows the user to simulate different inductor topologies, for a given working frequency and also to draw the inductance and quality factor curve along the entire frequency range for which the models are built.

The second tab in Fig. 3.23b, *Inductor Optimization*, allows the user to perform single- and multi-objective optimization of inductors. In single-objective optimization, the objectives of the optimization can be changed. The tool allows the user to maximize the quality factor, minimize the area, or both (by means of a weighted function where the weights of the function can be selected by the user) while achieving a given inductance. The multi-objective optimization can be performed either with two different objectives: a two-objective optimization, maximizing quality factor and inductance, or a three-objective optimization, where quality factor and inductance are maximized and area is minimized. In both optimization algorithms, constraints are applied in order to guarantee that the selected inductors can operate at the chosen working frequency (WF). These constraints are specified in Eqs. (3.43)–(3.47).

The third tab in Fig. 3.23c, *Build Model*, allows the user to build his/her own models for other technology processes or inductor topologies, by providing a training set for the surrogate modeling. By providing a test set it is also possible to validate the model automatically with the tool. Afterwards, the mean relative errors of the model are shown in a table for immediate model validation. Moreover, once the new model is built, it is automatically included as an option in the topologies popup menu of the other tabs and suitable for immediate simulation and optimization. All tabs have a message board and a README file, which are appropriate for an easy tool usage. The tool allows the user to generate an S-parameter file for any given inductor. This S-parameter file can afterwards be used in a modern simulator, such as HspiceRF or SpectreRF for an accurate description of the inductor behavior. For example, the AnalogLib library of Cadence has a device particularly intended for this purpose, the *nport*.

The SIDe-O tool, presented is this section, will be further used in this book in order to model inductors in several optimization-based circuit design methodologies.

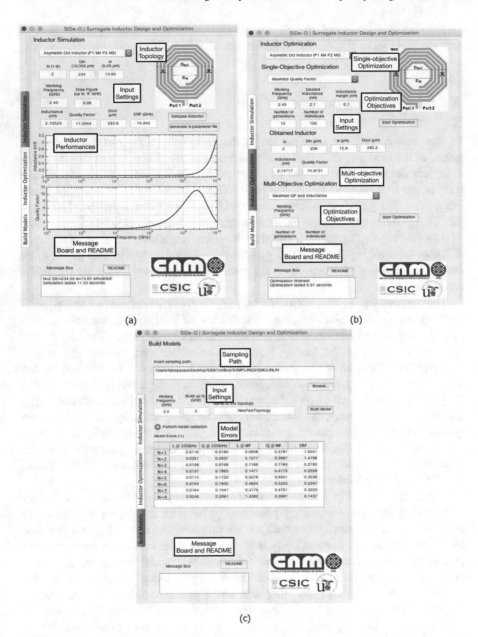

Fig. 3.23 SIDe-O graphical user interface (GUI) shown in separate tabs (Reprinted with permission from [35]). (**a**) Tab for inductor simulation. (**b**) Tab for inductor optimization. (**c**) Tab to build new inductor models

3.5 Summary

This chapter has discussed the design and modeling of integrated inductors. The chapter has given an insight on the geometric parameters of inductors and its performance parameters. Afterwards, an analytical model for integrated inductors was presented based on the segmented model. Moreover, such analytical model was compared against a surrogate model developed using machine learning techniques. The surrogate model has shown a far superior accuracy when compared to the analytical model. Moreover the models have been used for single- and multi-objective optimizations in order to design optimal inductors. In the end, a MATLAB toolbox has been developed which incorporates the ability to design and optimize inductors in a GUI in order to ease its usage by the RF designer.

References

1. K. Okada, K. Masu, *Modeling of Spiral Inductors* (INTECH Open Access Publisher, 2010)
2. C. Yue, S. Wong, Physical modeling of spiral inductors on silicon. IEEE Trans. Electron Devices **47**, 560–568 (2000)
3. N. Talwalkar, C. Yue, S. Wong, Analysis and synthesis of on-chip spiral inductors. IEEE Trans. Electron Devices **52**(2), 176–182 (2005)
4. S.S. Mohan, M.D.M. Hershenson, S.P. Boyd, T.H. Lee, Simple accurate expressions for planar spiral inductances. IEEE J. Solid State Circuits **34**, 1419–1424 (1999)
5. H. Greenhouse, Design of planar rectangular microelectronic inductors. IEEE Trans. Parts Hybrids Packag. **10**, 101–109 (1974)
6. Y. Koutsoyannopoulos, Y. Papananos, Systematic analysis and modeling of integrated inductors and transformers in RF IC design. IEEE Trans. Circuits Syst. II, Analog Digit. Signal Process. **47**, 699–713 (2000)
7. F. Passos, M. Kotti, R. González-Echevarría, M.H. Fino, M. Fakhfakh, E. Roca, R. Castro-López, F.V. Fernández, Physical vs. surrogate models of passive RF devices, in *IEEE International Symposium on Circuits and Systems* (2015), pp. 117–120
8. C. Wu, C. Tang, S. Liu, Analysis of on-chip spiral inductors using the distributed capacitance model. IEEE J. Solid State Circuits **38**, 1040–1044 (2003)
9. H. Hasegawa, M. Furukawa, H. Yanai, Properties of microstrip line on Si-SiO$_2$ system. IEEE Trans. Microwave Theory Tech. **19**(11), 869–881 (1971)
10. C. Wang, H. Liao, C. Li, R. Huang, W. Wong, X. Zhang, Y. Wang, A wideband predictive double-pi equivalent-circuit model for on-chip spiral inductors. IEEE Trans. Electron Devices **56**, 609–619 (2009)
11. F.W. Grover, *Inductance Calculations: Working Formulas and Tables* (Courier Dover Publications, 1929)
12. I. Bahl, *Lumped Elements for RF and Microwave Circuits* (Artech House, Boston, 2003)
13. F. Passos, M. Fino, E. Roca, R. Gonzalez-Echevarria, F. Fernández, Lumped element model for arbitrarily shaped integrated inductors—a statistical analysis, in *IEEE International Conference on Microwaves, Communications, Antennas and Electronics Systems* (2013), pp. 1–5
14. A.I.J. Forrester, A. Sobester, A.J. Keane, *Engineering Design via Surrogate Modelling: A Practical Guide* (Wiley, London, 2008)
15. ADS Momentum. http://www.keysight.com/en/pc-1887116/momentum-3d-planar-em-simulator. Accessed 1 Feb 2020

16. R. González-Echevarría, R. Castro-López, E. Roca, F.V. Fernández, J. Sieiro, N. Vidal, J. López-Villegas, Automated generation of the optimal performance trade-offs of integrated inductors. IEEE Trans. Comput. Aided Des. Integr. Circuits Syst. **33**, 1269–1273 (2014)
17. M.B. Yelten, T. Zhu, S. Koziel, P.D. Franzon, M.B. Steer, Demystifying surrogate modeling for circuits and systems. IEEE Circuits Syst. Mag. **12**(1), 45–63 (2012)
18. D. Gorissen et al., A surrogate modeling and adaptive sampling toolbox for computer based design. J. Mach. Learn. Res. **11**, 2051–2055 (2010)
19. DACE. http://www2.imm.dtu.dk/pubdb/views/publication_details.php?id=1460. Accessed 1 Feb 2020
20. MATLAB. https://www.mathworks.com/products/matlab.html. Accessed 1 Feb 2020
21. B. Bischl, O. Mersmann, H. Trautmann, C. Weihs, Resampling methods for meta-model validation with recommendations for evolutionary computation. Evol. Comput. **20**, 249–275 (2012)
22. F. Ferranti, L. Knockaert, T. Dhaene, G. Antonini, Parametric macromodeling based on amplitude and frequency scaled systems with guaranteed passivity. Int. J. Numer. Modell. Electron. Networks Devices Fields **25**(2), 139–151 (2012)
23. F. Passos, Y. Ye, D. Spina, E. Roca, R. Castro-López, T. Dhaene, F.V. Fernández, Parametric macromodeling of integrated inductors for RF circuit design. Microw. Opt. Technol. Lett. **59**(5), 1207–1212 (2017)
24. F. Passos, E. Roca, R. Castro-López, F.V. Fernández, Radio-frequency inductor synthesis using evolutionary computation and Gaussian-process surrogate modeling. Appl. Soft Comput. **60**, 495–507 (2017)
25. J. Kennedy, R. Eberhart, Particle swarm optimization, in *Proceedings of the IEEE International Conference on Neural Networks*, vol. 4 (1995), pp. 1942–1948
26. K. Deb, A. Pratap, S. Agarwal, T. Meyarivan, A fast and elitist multiobjective genetic algorithm: NSGA-II. IEEE Trans. Evol. Comput. **6**, 182–197 (2002)
27. K. Deb, An efficient constraint handling method for genetic algorithms. Comput. Methods Appl. Mech. Eng. **186**(24), 311–338 (2000)
28. S.K. Mandal, S. Sural, A. Patra, ANN- and PSO-based synthesis of on-chip spiral inductors for RF ICs. IEEE Trans. Comput. Aided Des. Integr. Circuits Syst. **27**, 188–192 (2008)
29. B. Liu, G.G.E. Gielen, F.V. Fernández, *Automated Design of Analog and High-frequency Circuits: A Computational Intelligence Approach, Studies in Computational Intelligence* (Springer, Berlin, 2014)
30. R. Gonzalez-Echevarria, E. Roca, R. Castro-López, F.V. Fernández, J. Sieiro, J.M. López-Villegas, N. Vidal, An automated design methodology of RF circuits by using Pareto-optimal fronts of EM-simulated inductors. IEEE Trans. Comput. Aided Des. Integr. Circuits Syst. **36**, 15–26 (2017)
31. E. Zitzler, L. Thiele, Multiobjective evolutionary algorithms: a comparative case study and the strength Pareto approach. IEEE Trans. Evol. Comput. **3**, 257–271 (1999)
32. G.G.E. Gielen, CAD tools for embedded analogue circuits in mixed-signal integrated systems on chip. IEEE Comput. Digital Tech. **152**, 317–332 (2005)
33. A.M. Niknejad, R.G. Meyer, Analysis, design, and optimization of spiral inductors and transformers for Si RF ICs. IEEE J. Solid State Circuits **33**, 1470–1481 (1998)
34. VeloceRF. https://www.ansys.com/products/semiconductors/ansys-velocerf. Accessed 28 Jan 2020
35. F. Passos, E. Roca, R. Castro-López, F.V. Fernández, An inductor modeling and optimization toolbox for RF circuit design, in Integration **58**, 463–472 (2017). https://doi.org/10.1016/j.vlsi.2017.01.009

Chapter 4
Systematic Design Methodologies for RF Blocks

In the previous chapter, the accuracy of two different inductor modeling techniques was assessed. It was concluded that surrogate models present higher accuracy when compared to analytical models. However, the influence of such accuracy on circuit performances was not yet studied. Therefore, the impact of such passive modeling techniques in systematic circuit design methodologies, must be examined. This is performed in Sect. 4.1, where an LNA design case study is adopted, in order to study the impact of using different inductor models on the overall circuit performances. Both the physical and the surrogate models will be integrated in optimization-based approaches in order to design the inductors presented in the adopted LNA topology. In Sect. 4.2, a hierarchical bottom-up design methodology is used in order to design the same LNA as in the previous section. Afterwards, performance and efficiency comparisons are performed between an optimization-based methodology which does not consider hierarchical decomposition versus a bottom-up design methodology which considers hierarchical decomposition. Therefore, it will be possible to conclude which is the most efficient methodology and, also, which one obtains superior results. In Sect. 4.3, the design of a VCO is considered in order to illustrate the problematic issue of process variability and the impact of not considering it within optimization-based methodologies. Finally, Sect. 4.4 illustrates the usage of optimization-based methodologies to design the remaining circuit of the front-end, the mixer. The mixer is a more complex circuit to design using optimization-based methodologies. On the one hand, the mixer testbench is more complex to set due to the RF and LO input signals and, on the other hand, the required analysis in order to get its performance parameters is time-consuming. Moreover, in all sections, and for all three RF blocks designed throughout the chapter, the different analysis performed will be presented and the strategies followed to improve the total optimization efficiency will be explained.

© Springer Nature Switzerland AG 2020

F. Passos et al., *Automated Hierarchical Synthesis of Radio-Frequency Integrated Circuits and Systems*, https://doi.org/10.1007/978-3-030-47247-4_4

4.1   The Impact of Passive Modeling Techniques in Systematic Circuit Design Methodologies: LNA Case Study

In this section, an LNA will be designed following a systematic circuit design methodology. The methodology used in this section is classified as a simulation-based sizing methodology because an electrical simulator will be used to estimate the performances of the LNA. Nevertheless, one of the most important issues of RF simulation-based circuit design methodologies is how passive components are designed, especially passive components that are highly affected by parasitics, such as integrated inductors. It was shown in Chap. 3 that physical models present poorer accuracy when compared to surrogate models; however, its impact on the design of RF circuits has not been studied yet. Therefore, this section will also focus on the usage of different inductor models. The objective is to show how can inaccurate inductor models impact the performances of RF circuits. The general optimization methodology used for this purpose as well as the user inputs is described in Fig. 4.1. The optimization algorithm selects the circuit sizing (active and passive components) and also the inductor geometric parameters at each iteration. The inductor model (e.g., surrogate or physical) then estimates the inductor behavior and the circuit is simulated by the RF circuit simulator (e.g., SpectreRF). The user inputs are a parameterized netlist, the passive and active device ranges, the objective(s) and constraints of the optimization and the number of solutions and iterations that the optimization should be performed with.

The difference between both methodologies is how the inductor is modeled. For the physical model, the π-model shown in Fig. 1.10 is used. This model has analytical equations that relate the geometrical parameters of the inductor (N, D_{in}, s and w) with a set of electrical components that try to emulate the physical behavior of the inductor. The equations of the model provide the value for each of its components (L_s, R_s, C_{ox}, etc.), and then the model can be instantiated in any electrical simulator in order to emulate an inductor. This is illustrated in Fig. 4.2 for a typical LNA topology.

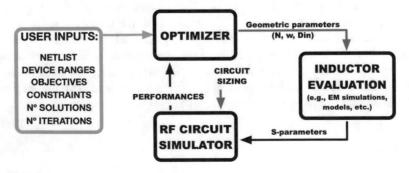

Fig. 4.1 General optimization-based methodology

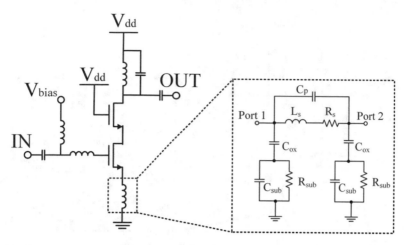

Fig. 4.2 Illustrating how the inductor is modeled with the π-model

On the other hand, the surrogate model does not provide any kind of relation between geometrical parameters and electrical components. Therefore, another way of using the surrogate model in electrical simulations must be used. The surrogate model presented in Chap. 3 was developed in order to estimate the performance parameters of the inductors (namely, L and Q), because the objective was to perform inductor synthesis. Therefore, the most straightforward option was to model those performance parameters which were then going to be directly optimized. In this chapter, the objective is to simulate the LNA, and, therefore, these L and Q values cannot be directly used in an electrical simulator in order to emulate the inductor behavior. Knowing that modern electrical simulators, such as SpectreRF, have electrical elements (the so-called *nport* devices) that use S-Parameter files in order to describe the behavior of any component, in this chapter, the strategy to model inductors L and Q presented in Chap. 3 is applied in order to model the inductor S-Parameters. The usage of the *nport* device and, consequently, integration of the S-Parameter files in the LNA simulation is illustrated in Fig. 4.3.

The surrogate modeling strategy is the same as the one presented in Chap. 3, where a model is created for inductors with different number of turns, and an inductor filtering is performed based on the inductors' SRF (see Fig. 4.4 for the sake of illustration). The training and validation inductor set was the same described in Chap. 3 (LHS sampling with 800 training inductors and 240 test inductors). Since a new model was created, a new error assessment is performed for each S-Parameter component (real and imaginary parts of S_{11}, S_{12}, S_{21}, and S_{22}) and shown in Table 4.1. The mean relative errors for each of the surrogate models are presented in Table 4.1. The errors might be perceived as relatively high errors (e.g., 1.89%). However, the comparatively larger percentage errors are due to the small values of some S-parameter components, which translates into relatively high percentages. Nevertheless, when all S-parameter components are combined in order to get L and

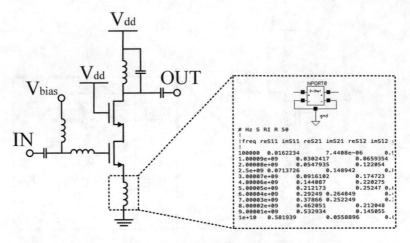

Fig. 4.3 Illustrating the nport device and how the inductor is modeled with the surrogate model

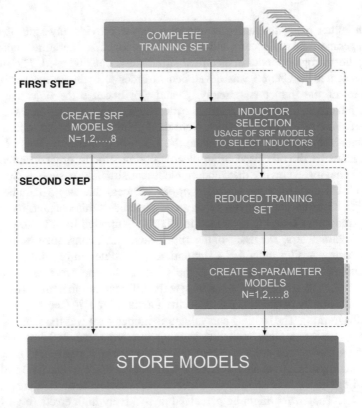

Fig. 4.4 Surrogate model construction strategy

Table 4.1 Mean relative error in % for the predicted values of S-Parameters with respect to EM simulations

N	Re S_{11}	Im S_{11}	Re S_{21}	Im S_{21}	Re S_{12}	Im S_{12}	Re S_{22}	Im S_{22}	L	Q
1	1.61	0.88	0.00	0.04	0.00	0.04	1.37	0.85	0.09	0.38
2	1.54	0.96	0.01	0.05	0.01	0.05	1.20	0.76	0.11	0.55
3	0.46	0.14	0.02	0.05	0.02	0.05	0.43	0.14	0.11	0.21
4	0.34	0.09	0.05	0.05	0.05	0.05	0.27	0.10	0.12	0.34
5	0.17	0.09	0.66	0.05	0.63	0.05	0.16	0.09	0.26	0.38
6	0.16	0.08	1.00	0.04	1.04	0.04	0.14	0.07	0.40	0.79
7	0.06	0.03	0.21	0.02	0.19	0.02	0.06	0.05	0.44	0.44
8	0.34	0.12	1.89	0.07	1.89	0.07	0.31	0.14	0.24	0.71

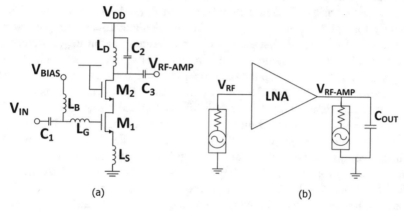

(a)   (b)

Fig. 4.5 (a) Schematic of the LNA and (b) its testbench

Q of a given inductor (as denoted in Sect. 3.1), this model shows errors always below 1%, as the last two columns in Table 4.1 show.

4.1.1   Experimental Results: Single-Objective Optimizations

The LNA topology used in the optimizations is the one presented in Fig. 4.5a and the testbench used for simulation is shown in Fig. 4.5b.

  The optimization process and the methodology itself is completely independent of the fabrication technology and the LNA topology used and only requires the user inputs previously mentioned (netlist, optimization objectives and constraints, circuit design variables, number of solutions and iterations). Regarding the circuit performances, the RF electrical simulator SpectreRF is used to evaluate each LNA. The most important performance parameters of the LNA were described in Chap. 2. The power consumption P_{DC} is extracted from a DC analysis. An S-parameter analysis provides the gain, input, and output matching, whereas a noise analysis provides the NF of the LNA. The area occupation is not accurately known until the circuit layout is performed. However, it is realistic to think that the layout area will be proportional to the sum of the areas of the individual devices (e.g., transistors, inductor, and capacitors in Fig. 4.5). Therefore, the use of this sum as estimation of the circuit area is a valid approximation for comparison of different candidate solutions.

The inclusion of IIP_3 in automated design methodologies usually takes long computation times since the IIP_3 calculation involves an input power sweep. Additionally, the IIP_3 calculation is difficult since linearity severely deteriorates with higher values of the input power. Therefore, selecting the best input power points to determine IIP_3 is not trivial and varies for each sized circuit. The method used in this work in order to efficiently include IIP_3 in the optimization process is based on the fact that IIP_3 is directly related with P_{DC} [1]. Therefore, once P_{DC} is calculated for a given sized circuit, IIP_3 can be calculated for an extrapolated input power well below P_{DC} so that the LNA is in the linear region. It was determined that 60 dB below P_{DC} guarantees a linear relation between input and output power. Using this method, the computation time for IIP_3 calculation is considerably reduced and it is feasible to include it within the LNA optimization loop.

In a first experimental test, single-objective optimizations are performed. Similarly to single-objective optimizations in Chap. 3, in this chapter, the PSO algorithm is used. Two optimizations were performed with 1000 iterations and 40 particles, one using the surrogate model to evaluate the inductor S-parameters and another one using the inductor π-model. Regarding the LNA design variables, these are shown in Tables 4.2 and 4.3. In Table 4.2, w_{Mi} are the gate widths of the i-th transistor, l_{Mi} are the transistor channel lengths, V_b is the bias voltage, and C_i represent the three-square capacitances used in the topology. In Table 4.3, the design variables for each inductor are shown (the number of turns N, the turn width W, and the inner diameter D_{in}).

Table 4.2 Design variables of the LNA optimization (other than inductors)

Design variables	Minimum	Maximum	Grid
w_{M1}, w_{M2} (μm) [a]	10	200	10
l_{M1}, l_{M2} (μm)	fixed @ 0.35 μm		
V_b (V)	0	1.5	0.001
C_1, C_2, C_3 (μm)	10	76	1

[a] Since the width of the foundry transistor model is only valid up to 200 μm, three parallel transistors are used for both M_1 and M_2 in order to increase the ranges. The min and max value shown are for each transistor

Table 4.3 Inductor design variables

Parameter	Minimum	Grid	Maximum
N	1	1	8
D_{in} (μm)	10	1	300
W (μm)	5	0.05	25
s (μm)	2.5	-	2.5

The optimization was performed with the objectives and constraints shown in the second column of Table 4.4. The LNA performances obtained with both methods can be seen in the third and sixth columns in Table 4.4 and the values of the circuit elements are shown in Table 4.5. By using these models instead of EM simulation to evaluate the inductor performances, some error in the LNA performances can be expected due to the limited model accuracy. Therefore, the inductors obtained from the LNA optimization (with both methods) were simulated electromagnetically and the simulations of the final LNAs were performed again using this more accurate evaluation of the inductor performances. The performance shifts can be observed in the fourth and seventh columns in Table 4.4. It can be observed that the LNA performance shifts are negligible (less than 1%) when the surrogate model developed herein is used. However, when the analytical π-model is used, huge shifts are observed (up to 65% in this experiment). In some cases, by using the π-model, some of the design constraints are no longer met when the inductors are electromagnetically simulated and, therefore, the design is not valid (e.g., NF in column 7 in Table 4.4).

As the PSO algorithm is stochastic, different executions may yield different results; therefore, the algorithm was executed five times. The statistical analysis for the design objective can be observed in Table 4.6. When the results of all these executions are simulated again with electromagnetically simulated inductors, it is found that design constraints were violated four times when the π-model was used for the optimization, whereas there is no violation when the surrogate model was used. For the sake of illustration, graphical plots were performed with the circuit simulator SpectreRF for the LNAs obtained using the surrogate model of the inductors. Two simulations were performed: one using the S-parameters of the inductors obtained by the models and another with the S-parameters of the same inductors but electromagnetically simulated. Figure 4.6 shows the comparisons for the gain, the input and output matching, and the noise figure. Again, it is possible to realize that the inductor surrogate model is highly accurate since the performance curves of the LNA are completely overlapped.

The time needed for the LNA optimization with PSO in a two 6-core Intel Xeon E5-2630 v2 processors at 2.60 GHz is 0.30 h and 3.75 h CPU time for the example using the π-model and the surrogate model, respectively. By using the analytical model a greater efficiency is obtained; however, it is largely disadvantageous due to the huge performance shifts obtained when this model is used. As discussed above, the accuracy using the surrogate model is comparable to using EM simulations of the inductors. However, if we take into account that the EM simulation of a single inductor typically takes from a few minutes to several hours (let us assume a conservative average of 10 min per inductor), this means that the optimization process described above would need 9 months ($40 \times 1000 \times 10$ min) only to electromagnetically evaluate the inductor performances (obviously, the elapsed time can be decreased by using parallelization). Therefore, the relevance of obtaining practically the same results in less than 4 h can be understood.

Table 4.4 Objectives and constraints for the LNA optimization with PSO. Performances obtained for the LNA optimization using the surrogate model and the π-model. Performance deviation when inductors are EM simulated

Performance	Constraint /Objective	Performances using surrogate model	Performances using surrogate model (after EM)	Performance deviation (%)	Performances π-model using π-model	Performances using π-model (after EM)	Performance deviation (%)
S_{11}	<-10 dB	-13.951 dB	-13.990 dB	0.145	-10.983 dB	-11.762 dB	6.621
S_{22}	<-10 dB	-23.887 dB	-23.828 dB	0.247	-28.517 dB	-17.314 dB	64.705
S_{21}	>15 dB	15.168 dB	15.164 dB	0.026	15.214 dB	15.010 dB	1.360
K	>1	14.056	14.077	0.149	12.32	15.86	22.320
NF	<3 dB	2.855 dB	2.859 dB	0.158	2.940 dB	3.178[a]	7.477
P_{DC}	<10 mW	9.999 mW	9.999 mW	0	9.999 mW	9.999 mW	0
IP_3	>-10 dBm	-1.353 dBm	-1.364 dBm	0.785	-3.162 dBm	-2.998 dBm	5.461
Inductors	Well-behaved @ 2.4–2.5 GHz[b]	yes	yes	–	yes	yes	–
Area (μ m^2)	Minimize	7.88×10^4	7.88×10^4	–	8.39×10^4	8.39×10^4	–

[a] This performance fails to meet the constraint
[b] Well-behaved means that the inductors comply with the constraints defined in Chap. 3

Table 4.5 Design variables for the LNA designs obtained from optimization with PSO using both inductor models

Inductor Model	W_1 (μm)	W_2 (μm)	$l_{1,2}$ (μm)	V_b (V)	C_1 (pF)	C_2 (pF)	C_3 (pF)	L_S	L_G	L_B	L_D
Surrogate	490	310	0.35	0.783	0.622	1.199	0.550	$N=1$ $D_{in} = 91\,\mu m$ $W=11.30\,\mu m$	$N=1$ $D_{in} = 49\,\mu m$ $W=5.25\,\mu m$	$N=5$ $D_{in} = 51\,\mu m$ $W=6.40\,\mu m$	$N=3$ $D_{in} = 155\,\mu m$ $W=6.30\,\mu m$
π-model	600	495	0.35	0.753	0.826	1.977	0.947	$N=1$ $D_{in}=10\,\mu m$ $W=12.90\,\mu m$	$N=5$ $D_{in}=21\,\mu m$ $W=8.50\,\mu m$	$N=3$ $D_{in}=113\,\mu m$ $W=9.95\,\mu m$	$N=3$ $D_{in}=105\,\mu m$ $W=9.05\,\mu m$

Table 4.6 Statistical results of the LNA area obtained in different runs of the single-objective optimization using PSO

Inductor model	Mean (μm^2)	Best (μm^2)	Worst (μm^2)	Constraint violation
Surrogate model	1.17×10^5	7.88×10^4	1.39×10^5	0/5
π-model	1.35×10^5	8.39×10^4	2.29×10^5	4/5

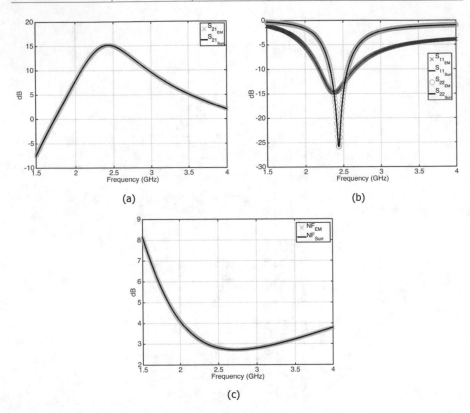

Fig. 4.6 Comparisons of the LNA performances using the inductors obtained with the surrogate model and the same inductors electromagnetically simulated. (**a**) Gain (S_{21}), (**b**) Input (S_{11}) and output (S_{22}) matching, and (**c**) Noise figure (NF). (Reprinted with permission [2])

4.1.2   Experimental Results: Multi-Objective Optimizations

It has been seen in the previous section that the usage of the surrogate model is very advantageous regarding the performance accuracy. Nevertheless, another experiment is performed using multi-objective optimization algorithms (the NSGA-II algorithm is used) in order to further observe the advantages that appear due to the usage of a surrogate model over a physical model. Therefore, two multi-objective optimizations were performed: one with the π-model as inductor evaluator and another with the surrogate model. The optimization objectives were the maximization of S_{21} and the minimization of noise figure and power consumption. The design specifications and constraints for this optimization are given in Table 4.7. The optimization was performed with 1000 individuals and 300 generations. The comparison between the obtained POFs is shown in Fig. 4.7.

Furthermore, it is expected that a shift in the LNA performances may occur due to the usage of the models. Therefore, the inductors used in the LNAs were EM simulated and the LNAs were re-simulated in order to observe how the performances shift. In Figs. 4.8 and 4.9 it is possible to observe the new LNA POFs with the inductors EM simulated for the surrogate model and the π-model, respectively. It can be observed in Fig. 4.8 that the shifts in the POF obtained with the surrogate model are negligible, whereas the shifts obtained when the π-model is used are much more noticeable (see Fig. 4.9). Furthermore, all the LNAs obtained using the surrogate model still meet the design constraints after the inductors are

Table 4.7 Desired performances and constraints for the LNA optimization with objectives Gain vs. NF vs. Power

LNA performances	LNA specifications
S_{11}	$<-10\,\mathrm{dB}$
S_{22}	$<-10\,\mathrm{dB}$
S_{21}	Maximize
S_{21}	$>10\,\mathrm{dB}$
k	>1
IIP_3	$>-10\,\mathrm{dBm}$
NF	Minimize
NF	$<4.5\,\mathrm{dB}$
Area $\mu\ \mathrm{m}^2$	$<1 \times 10^5\ \mu\mathrm{m}^2$
P_{DC}	Minimize

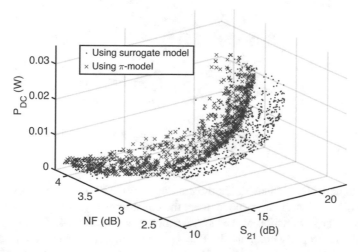

Fig. 4.7 POF comparison of the optimization Gain vs. NF vs. Power using the surrogate and the π-model (Reprinted with permission [3])

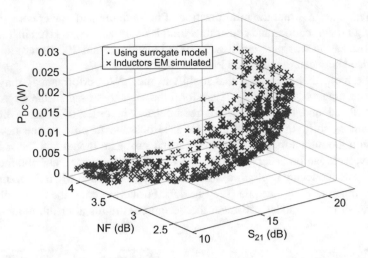

Fig. 4.8 POF comparison of the optimization Gain vs. NF. vs. Power using the surrogate model and after its inductors have been EM simulated (all LNAs meet constraints after the inductors being EM simulated) (Reprinted with permission [3])

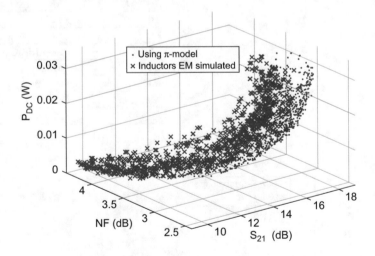

Fig. 4.9 POF of the optimization Gain vs NF vs Power using the π-model and after its inductors have been EM simulated (Reprinted with permission [3])

electromagnetically simulated. On the contrary, only 129 out of the 1000 LNAs in the POF obtained with the π-model meet the constraints after EM simulation of the inductors and re-evaluation of the LNA performances (see Fig. 4.10).

In order to properly compare the POFs, some comparison metrics must be used and a statistical study should be performed. Among the several metrics available in the literature, coverage set and hypervolume will be used (the same metrics used to compare POFs in Chap. 3). Since the optimization times are relatively high,

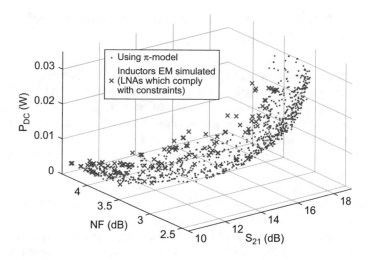

Fig. 4.10 POF of the optimization Gain vs NF vs Power using the π-model (only LNAs which meet constraints after the inductors being EM simulated are depicted) (Reprinted with permission [3])

Table 4.8 Statistical results of the front metrics obtained in different runs using the surrogate model and the π-model

Metric	Mean	Max	Min
Hypervolume surrogate model	0.414	0.481	0.298
Hypervolume π-model	0.262	0.365	0.213
$C(\pi, \text{surrogate})$	0.968	1	0.839
$C(\pi, \text{surrogate})$	0.014	0.067	0

the number of runs was limited to five. Table 4.8 shows the statistical results of hypervolume of the fronts obtained using the surrogate model and the π-model after EM simulation of the final inductors, so that the accuracy of the simulations is comparable. It can be observed that the hypervolume using the surrogate model is consistently larger. Furthermore, Table 4.8 also shows the coverage set between the couple of fronts: that obtained with the π-model for the inductor and that obtained with the surrogate model. It can be observed that most solutions of the POF obtained with the π-model are dominated by the POF obtained with the surrogate model, whereas very few solutions using the surrogate model are dominated by those using the π-model. The time needed for the LNA optimization with NSGA-II in a two 6-core Intel Xeon E5-2630 v2 processors at 2.60 GHz is 31.93 h CPU time using the surrogate model and 6.05 h CPU time using the π-model. As in PSO, a greater efficiency is obtained by using the π-model; however, huge performance shifts are obtained. Nevertheless, the surrogate model provides an excellent accuracy comparable to EM simulation, whereas the latter one would be computationally unaffordable.

4.2   LNA Design Using a Bottom-Up Systematic Circuit Design Methodology

In this section, a different optimization-based sizing methodology based on a bottom-up design strategy is going to be applied to the design of an LNA. The basic idea is to start the design process at the lowest possible level: passive component level, i.e., to bring the hierarchical partitioning and bottom-up hierarchical composition design paradigm down to the device level [4]. Since resistors and capacitors do not have such convoluted trade-offs as inductors, only the latter is considered for a POF generation.

This methodology is based on the assumption that if the designer generates an inductor POF before the circuit optimization, the best inductor designs will be available in the POF and that no other inductor would be capable of improving the circuit performances. The previously proposed surrogate model has proven to be an attractive inductor modeling solution which can accurately compute the behavior of inductors while being computationally cheap to evaluate. Therefore, in this section the capabilities of the previously presented surrogate model will be used in order to efficiently and accurately evaluate inductors. By doing so, it is possible to assess which is the most advantageous RF design methodology: the optimization-based methodology used in the previous section, where the inductors are selected during the circuit optimization stage (again illustrated in Fig. 4.11 for convenience) or a bottom-up optimization-based methodology, where the inductors are optimized before any circuit optimization (see Fig. 4.12). The idea illustrated in Fig. 4.12 is that the designer first runs a multi-objective optimization in order to achieve an inductor POF (as performed in Sect. 3.3.2), and then this POF is used as an optimal inductor library during the circuit optimization.

For the sake of simplicity, from now on, the optimization-based methodology where the inductor geometrical parameters are chosen during the optimization will be denoted as online strategy (because the inductors are designed online during the circuit optimization) and the bottom-up optimization-based methodology will be denoted as offline (because the inductors are designed offline, prior to any circuit optimization). In this chapter both strategies are applied to the design of the LNA in Fig. 4.5 in order to inspect which strategy shows better performance and efficiency

Fig. 4.11 Online bottom-up design strategy (Reprinted with permission [5])

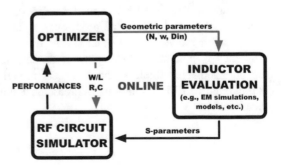

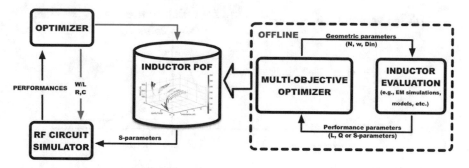

Fig. 4.12 Offline bottom-up design strategy (Reprinted with permission [5])

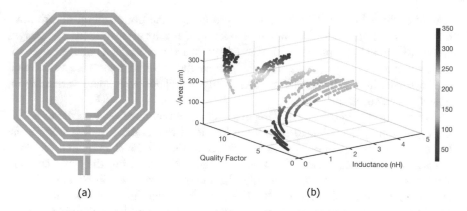

(a)   (b)

Fig. 4.13 Inductor topology and POF used in the LNA optimization. (**a**) Instance of the octagonal asymmetric inductor. (**b**) POF of 1000 points with the trade-off $\sqrt{\text{Area}}$ vs. Quality factor vs. Inductance of the octagonal asymmetric inductor

in the design of RF circuits. For the offline design strategy, an inductor POF has to be generated prior to any circuit optimization. For this purpose, the SIDe-O tool, presented in Sect. 3.4, can be used. A multi-objective optimization using 1000 individuals and 80 generations was performed, which took around 10 min to run in SIDe-O. The search space is presented in Table 4.3 (the same design space is used for both online and offline strategies). The optimization was performed with three objectives: maximize quality factor and inductance, at the frequency of interest, while minimizing the area. The inductors were subjected to some constraints used to guarantee that the inductors are in the flat-BW zone, as defined in Sect. 3.3.1, in Eqs. (3.43)–(3.47). The inductor topology used is illustrated in Fig. 4.13a and the obtained POF can be seen in Fig. 4.13b.

In high-level block optimization, the low-level POFs have to be explored because they represent part of the search space in the high-level optimization. Therefore, an important issue in bottom-up design strategies is how to search through these low-level POFs when optimizing high-level blocks. This issue is important because

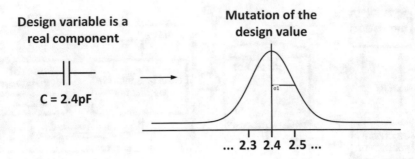

Fig. 4.14 Illustrating the mutation operation for regular design variables

searching these low-level POFs poses a problem to the optimization algorithms. Evolutionary algorithms use mutation operators for local search in the design space, where a slight movement in the design space represents a small change of element parameters, e.g., transistor width, capacitance value, resistance value, etc. (as illustrated in Fig. 4.14), that are commonly associated with small variations in objective and constraint values.

When considering low-level POFs in a high-level optimization, in order to allow the optimization algorithm to search the low-level POFs, the simplest solution is to assign an integer value to each individual of the low-level POF, and use this integer (the so-called index value) as a design variable during the optimization. The range of this new design variable, the index variable, would be the number of individuals in the low-level POF. The problem is that this index variable does not have any information on the performances of the low-level individual, and, therefore, individuals with index 1, 2, or 3 may be in completely different areas of the design space (see Fig. 4.15). Hence, while performing mutation around individual 1, the algorithm can jump to individual 3, which is in a completely different area of the design space. As a consequence, the mutation operation is transformed into an arbitrary random variation, which can hamper the convergence of the optimization.

In order to solve this problem, and, realizing that a POF generated for N design objectives is a hyper-surface of dimension N-1, a set of N-1 coordinates can be used to represent the POF. Therefore, instead of assigning only one index to the individuals in low-level POFs, the individuals of each POF can be sorted by their performances and mapped into a matrix with N-1 dimensions. In Fig. 4.16, it is possible to see this operation, where a set of two coordinates is given to each individual of a 3D POF. By doing so, each individual of a low-level POF can be represented by a set of coordinates instead of a single index variable. Thus, the designs in each POF are organized by their performances in such a way that the mutation operator can be efficiently used. The matrix coordinates are then used as design variables in the upper level optimization [6].

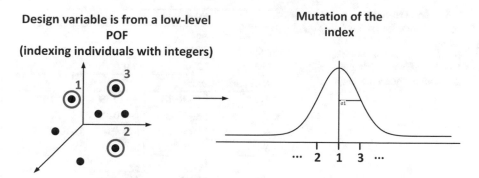

Fig. 4.15 Illustrating the mutation operation for indexed variables

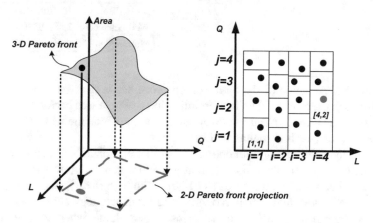

Fig. 4.16 Illustrating the mapping of each inductor into a two-coordinate matrix

4.2.1 Experimental Results: Multi-Objective Optimizations

In order to compare both strategies, three different optimizations were performed considering different objectives and constraints. For each experiment, the constraints, objectives, number of individuals and generations used can be observed in Table 4.9. The LNA design variables are shown in Tables 4.2 and 4.3. For the online design strategy, where the inductors are optimized online, the total number of design variables is 18 (variables in Table 4.2 plus three variables per inductor). For the offline strategy, where the inductors are selected from the inductors POF, the number of variables is reduced to 14 (variables in Table 4.2 plus two indexes per inductor). All optimizations in this section were performed for the LNA topology shown in Fig. 4.5 in a 0.35 μm-CMOS technology intended to operate at the frequency band of 2.4–2.5 GHz, with a supply voltage V_{DD}=2.5 V.

Table 4.9 Desired
Specifications for the LNA
optimization using NSGA-II
for three different examples.
Number of individuals and
generations for each example

	Example 1	Example 2	Example 3
Optimization settings			
Individuals	300	300	300
Generations	200	200	200
LNA performances			
S_{11}	$<-10\,dB$	$<-10\,dB$	$<-10\,dB$
S_{22}	$<-10\,dB$	$<-10\,dB$	$<-10\,dB$
S_{21}	Maximize	$>10\,dB$	$>10\,dB$
S_{21}	$>10\,dB$	$>10\,dB$	$>10\,dB$
k	>1	>1	>1
NF	$<3\,dB$	Minimize	$<3.5\,dB$
P_{DC}	$<25\,mW$	$<25\,mW$	Minimize
IIP_3	$>-10\,dBm$	$>-10\,dBm$	$>-10\,dBm$
Inductors	Well-behaved @ 2.4–2.5 GHz		
Area (μm^2)	Minimize	Minimize	Minimize

Since each execution of the optimization processes is stochastic, ten runs were made for each experiment. The first optimization example was performed with two objectives: maximize S_{21} and minimize area. The hypervolume was calculated for the twenty runs (for both online and offline methods) and the POFs with highest and lowest hypervolume of both online and offline optimizations[1] can be observed in Fig. 4.17a. It is possible to observe that the offline run with lowest hypervolume achieves higher S_{21} and lower areas (for LNAs with $S_{21}>18\,dB$), when compared to the online POF with higher hypervolume. The online strategy achieves far poorer results on the run which achieved the lowest hypervolume, when compared to the offline strategy.

The evolution of the hypervolume versus each generation of the optimization is illustrated in Fig. 4.17b for the online experiment and in Fig. 4.17c for the offline experiment (all ten executions can be observed). The hypervolume is only calculated when feasible solutions are available in the POF; therefore, in the first generations of the optimization, the hypervolume is zero. It is possible to conclude from the hypervolume plots that the hypervolumes are stable since early generations, which means that increasing the number of generations would not improve the quality of the POF. It is also possible to observe that the quality of the offline POFs is much more consistent than the online ones.

The second optimization example also has two objectives: minimize area and noise figure. The POFs can be seen in Fig. 4.18a. In this design example, the highest hypervolume POFs of both strategies are similar; however, it is possible to observe that the offline strategy allows relatively lower noise figures and lower areas. The offline POF is also wider, achieving lower noise figures, while sacrificing area.

---

[1]The same reference point is used for the hypervolume calculation of all offline and online experiments.

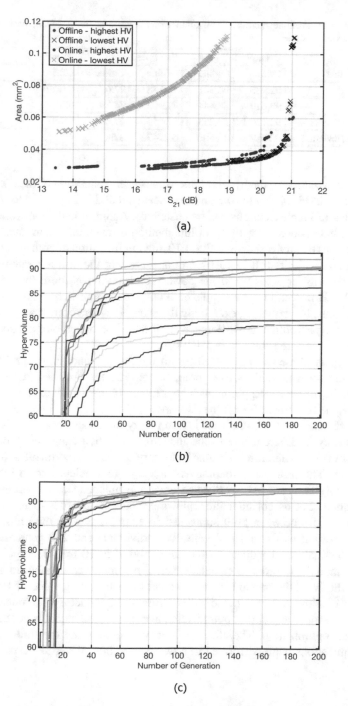

Fig. 4.17 (**a**) LNA POF obtained for the trade-off S_{21} vs. Area (example 1). From the ten runs of each experiment, the one with highest and lowest hypervolume were selected. (**b**) Illustrating the evolution of the hypervolume versus the number of generations of the optimization, for example 1 for the online experiment, and (**c**) the offline experiment

Table 4.10 Hypervolume metrics, for example, one, two, and three. Mean, best, worse, and standard deviation values

Hypervolume	Example 1		Example 2		Example 3	
	Offline	Online	Offline	Online	Offline	Online
Best	92.986	92.381	78.158	77.6	0.0760	0.0753
Mean	92.620	88.444	77.962	74.59	0.0754	0.0714
Worse	92.120	79.253	77.443	67.65	0.746	0.0643
Standard Deviation	0.2817	4.360	0.2147	3.68	4.18×10^{-4}	0.0038

Again, the evolution of the hypervolume versus each generation of the optimization is illustrated in Fig. 4.18b for the online experiment and in Fig. 4.18c for the offline experiment (all ten executions can be observed). Again, it is demonstrated that the quality of POFs obtained with the offline method is much more consistent.

In the third example, shown in Fig. 4.19, the minimization of both area and power consumption is desired. It is possible to observe that the offline strategy achieves designs with lower areas when the power consumption is more than 5 mW (see Fig. 4.19a). However, in this example both POFs with highest hypervolume achieve similar results. Nevertheless, examining the example where both strategies achieved the lowest hypervolume, the difference is vibrant. The hypervolume curves for the ten different runs are shown in Fig. 4.19b, c.

The hypervolume was calculated for all the 60 runs and the mean, best, worse results are shown in Table 4.10 in order to compare the obtained POFs. It can be seen that the offline strategy obtains higher hypervolume for all examples, proving, therefore, the benefits of the offline strategy.

The efficiency of both strategies is also a very important subject. It is clear that the efficiency is different due to the inductor evaluation stage being done either online or offline. Therefore, the time necessary for each experiment is depicted in Table 4.11. The offline methodology is much more efficient due to the fact that the inductors POF are only performed once and independently from each circuit optimization. Let us consider the following example in order to understand the time differences between both strategies: in example 3 with the online strategy, the optimization was performed with 300 individuals and 200 generations. Each LNA individual has 4 inductors (that are analyzed in 10 frequency points with 8 different models, in order to calculate each S-parameter component in Table 4.1). If we multiply all the previous values, we get the astonishing value of 19.2 million model evaluations, whereas in the offline strategy, the inductor S-parameter files are calculated prior to any circuit optimization and, hence, the model is not used during the circuit optimization. Therefore, it is easy to understand the time differences. All optimizations were performed in a two 6-core Intel E5-2639 v2 processors at 2.60 Hz.

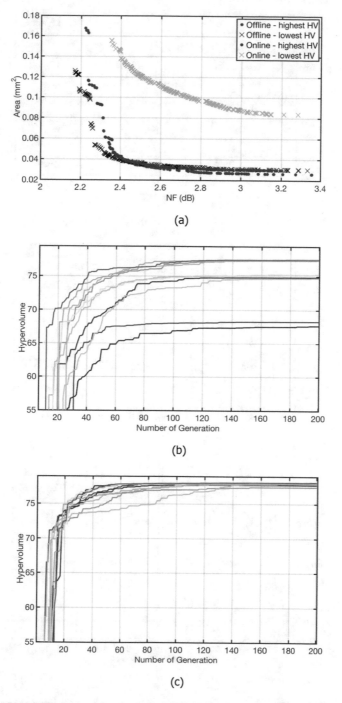

Fig. 4.18 (**a**) LNA POF obtained for the trade-off NF vs. Area (example 2). From the ten runs of each experiment, the one with highest and lowest hypervolume were selected. (**b**) Illustrating the evolution of the hypervolume versus the number of generations of the optimization, for example 2 for the online experiment, and (**c**) the offline experiment

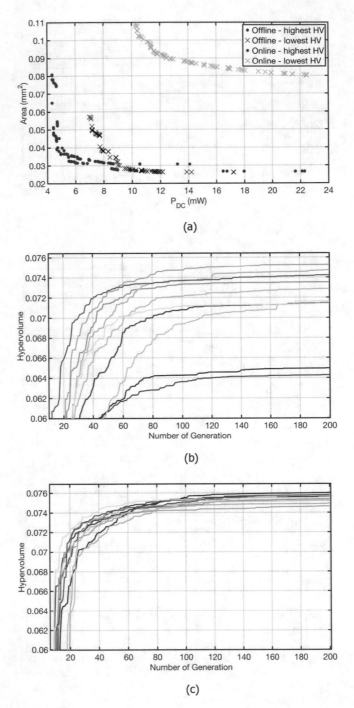

Fig. 4.19 (**a**) LNA POF obtained for the trade-off P_{DC} vs Area (example 3). From the ten runs of each experiment, the one with highest and lowest hypervolume were selected. (**b**) Illustrating the evolution of the hypervolume versus the number of generations of the optimization, for example 3 for the online experiment, and (**c**) the offline experiment

Table 4.11 CPU time for the different experimental examples

	Offline	Online
Experiments	Time (h)	Time (h)
Inductor POF	0.20	-
Example 1	1.23	6.15
Example 2	1.24	6.10
Example 3	1.49	6.15

4.3 VCO Design using a Bottom-up Systematic Circuit Design Methodology and Considering Process Variability

In the first sections of this chapter, two different issues of systematic RF circuit design methodologies were considered. In Sect. 4.1 the importance of accurate inductor models was illustrated, and in Sect. 4.2 the benefits of the enhanced optimization methodology, based on bottom-up approaches, were illustrated.

In this section another important issue for an accurate circuit design is tackled: the inclusion of process variability in systematic circuit design methodologies. In this section, an optimization-based approach is adopted to systematically design a voltage controlled oscillator (VCO), where the passive component design problem is tackled by using the state-of-the-art surrogate modeling technique previously presented in Chap. 3 and the enhanced circuit design methodology presented in 4.2. Furthermore, by considering process parameter variations during the optimization (e.g., extreme corner performances) for each candidate design, robust designs are achieved, which are closer to a first-pass fabrication success.

4.3.1 Experimental Results: Multi-Objective Optimizations

The VCO topology used in order to perform the experimental tests is shown in Fig. 4.20a and the testbench used for optimization is shown in Fig. 4.20b. From the several VCO topologies available, in this work, a cross-coupled double-differential VCO is considered. In order to have an oscillation, a negative resistance must be generated to compensate the parasitic resistance of the VCO tank (formed by the capacitor, C, and inductance, L). In the circuit presented in Fig. 4.20, both the PMOS and NMOS transistors are used to provide sufficient negative resistance to make the oscillator operate uninterruptedly. This LC-VCO topology shows good phase noise performance as well as low power consumption [7]. From the available voltage and current biasing techniques, the current one is selected in this work due to its lower power consumption [7].

In order for a VCO to oscillate properly, the designer has to guarantee that the negative resistance (given by the MOS transistors) is higher than the resistance imposed by the tank (mainly by the inductor). Therefore, an established rule-of-thumb is that the following condition, the so-called start-up condition, is reached:

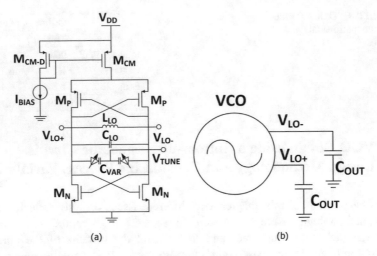

Fig. 4.20 (a) Schematic of the VCO. (b) VCO testbench

$$g_{active} \geq \alpha \times g_{tank} \tag{4.1}$$

where g_{active} and g_{tank} are the transconductances of the MOS transistors and the tank loss, respectively, and α is a constant value, often between 2 and 4. The oscillation frequency f_{osc} is given by

$$\frac{1}{\sqrt{LC_{\max}}} \leq f_{osc} \leq \frac{1}{\sqrt{LC_{\min}}} \tag{4.2}$$

where L is the inductance and $C_{\min}$ and $C_{\max}$ are the minimum and maximum tank capacitances, which vary due to the varactors.

Apart from f_{osc}, which only depends on the total capacitance and inductance of the circuit, and, therefore, does not need special design techniques, other performances, such as phase noise and power consumption, are usually conflicting, making the design of the VCO extremely complex [7]. Therefore, some design strategies are required which demand some specific design procedure and a great deal of expert knowledge from the designer. One of the most common design strategies that usually fulfills both, phase noise and power specifications, can be summarized as follows. The designer must find the minimum inductance that satisfies both the start-up condition and the output swing. Then, the maximum bias current allowed by the design specifications is chosen. This will ensure the maximization of the output swing and the minimization of the phase noise. Furthermore, the tank capacitance also introduces a trade-off in the design, where large capacitances improve the phase noise but reduce the tuning range [8].

In summary, the manual knowledge-based design of such VCOs is not trivial and, frequently, re-design iterations are needed in order to achieve the desired specifications. Thus, in this section, an optimization algorithm that searches for optimal designs is considered. The first step in the enhanced optimization methodology is to generate the inductor POF. In order to do so, SIDe-O (presented in Sect. 3.4) is used. In Fig. 4.21a, an octagonal symmetric inductor is presented, which is later going to be used in the circuit under study in this section. The design variables of the inductors used to generate the POF are shown in Table 4.12. As in previous optimizations, the inductor area was limited to a maximum square of $400\,\mu m \times 400\,\mu m$. The POF was obtained from an optimization with a population of 1000 individuals and 80 generations, which took around 10 min to run (see Fig. 4.21b). After obtaining the inductor POF, the inductors are mapped into a matrix and defined by two index coordinates in order to allow its usage in the circuit optimization (as described in the previous section).

After obtaining the inductor POF, it is possible to proceed to the circuit optimization. The circuit design variables are shown in Table 4.13. In this specific topology, there are 15 initial design variables: the p-type and n-type MOS transistors sizes (w_n, w_p, l_n, and l_p), the indexing variables of the integrated inductor (two indexes for the matrix mapping indexing procedure), the varactor sizes (w_{var} and l_{var}), the capacitor value (C), the sizes of the transistors used to bias the circuit (w_d, w_{dd}, l_d and l_{dd}), and also, the bias current (I_{bias}). The optimization was performed using the testbench shown in Fig. 4.20b. The optimization objectives and constraints can be seen in Table 4.14. The circuit is intended to operate at Vdd=2.5 V. Three analysis need to be performed in order to measure all the desired performances: a periodic steady-state (PSS) analysis, in order to calculate the oscillation frequency

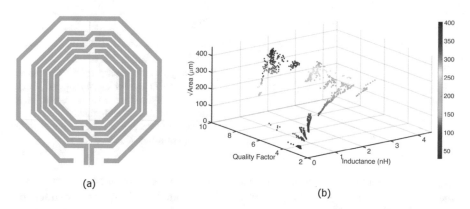

(a)

(b)

Fig. 4.21 Inductor topology and POF used in the VCO layout-aware optimization. (**a**) Instance of the octagonal symmetric inductor with guard ring. (**b**) POF of 1000 points with the trade-off $\sqrt{Area}$ vs. Quality factor vs. Inductance of the octagonal symmetric inductor with guard ring

Table 4.12 Design variables of the inductors

Parameter	Minimum	Grid	Maximum
N	1	1	8
D_{in} (μm)	10	1	300
w (μm)	5	0.05	25
s (μm)	2.5	-	2.5

Table 4.13 Design variables of the VCO

Variables	Minimum	Maximum	Grid
w_{n1} (μm)	10	200	10
$w_{p1,d,dd}$ (μm)	10	150	10
$l_{n1,p1,d,dd}$ (μm)	Fixed @ 0.35 μm		
I_{bp} (mA)	0.1	1.5	0.1
w_{Cvar} (μm)	Fixed @ 6.6 μm		
l_{Cvar} (μm)	Fixed @ 0.65 μm		
Inductors	Selected from the POF		
Row_{Cvar}, Col_{Cvar} [a]	4	12	1
C (μm) [b]	10	76	1

[a] Row_{Cvar} and Col_{Cvar} are the number of fingers per row and column of the varactors
[b] Two parallel capacitors are used. The min and max values are for each poly capacitor used

Table 4.14 Design specifications for the VCO optimization

VCO performance	VCO specifications
f_{osc} (V_{tune}=0 V) [a]	>2.55 GHz
f_{osc} (V_{tune}=2.5 V) [a]	<2.45 GHz
PN @ 10 kHz	<−65 (dBc/Hz)
PN @ 100 kHz	<−92 (dBc/Hz)
PN @ 1 MHz	Minimize
PN @ 1 MHz	<−113 (dBc/Hz)
PN @ 100 MHz	<−134 (dBc/Hz)
P_{DC}	Minimize
P_{DC}	<25 mW
V_{OUT}	>0.15 V
Area (μm^2) [b]	Minimize

[a] By applying these constraints we ensure that the VCO covers the target oscillation frequency of 2.5 GHz, ensuring at the same time an acceptable tuning range
[b] The area of each VCO is approximated as the sum of all device sizes

f_{osc} and the output voltage V_{OUT}, a steady-state noise analysis, in order to calculate the phase noise at different frequencies, and a DC analysis in order to extract the power consumption P_{DC}.

The optimization was performed with 256 individuals and 250 generations and the results are shown in Fig. 4.22. The entire optimization took approximately 5 h to run in an Intel Core i7-3770 @ 3.4 GHz workstation with 32 Gb of RAM. It is important to recall that SIDe-O uses a highly accurate surrogate modeling that already incorporates the inductor parasitics at high frequencies. Yet, some performance deviations could still occur due to the usage of the model. Therefore, as a verification test, all the inductors were EM simulated and the VCOs were re-simulated. Due to the fact that the surrogate model is so accurate, the results of

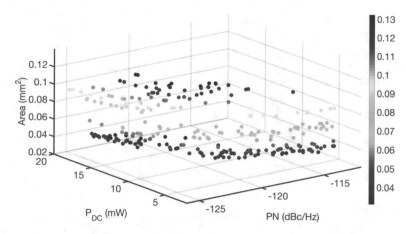

Fig. 4.22 VCO POF obtained from the enhanced two-step optimization considering only typical process parameters (Reprinted with permission [9])

this verification are that all individuals still comply with constraints and the VCO performances suffer minimal variations (at centesimal level).

However, some important aspects of the circuit design were still unconsidered during this optimization, such as the device corners. Therefore, in order to inspect how damaging could the corner evaluation be to the designs available in the POF, all the VCO designs were re-simulated considering their corners. The following corners were considered: the worst-case power condition (WP), the worst-case speed condition (WS), the worst-case one (WO), where the circuit is designed with fast NMOS and slow PMOS, and the so-called worst-case zero (WZ), where the circuit is designed with slow NMOS and fast PMOS.

It was found that from the 256 VCO designs obtained in the POF from Fig. 4.22, none of them complied with constraints after being evaluated at corner conditions, which would mean that none of the designs found in the sizing optimization would be valid in all its corners. In Fig. 4.23 it is possible to observe the PN vs. P_{DC} projection (of the POF obtained in Fig. 4.22 in blue crosses) and how the POF was shifted, in red crosses (the area is not affected by the corners). The corner performances depicted are the worst possible for each corner of the design. This means that for the VCO designs depicted with red crosses in Fig. 4.23 their performances may not be authentic, as the PN value may be due to one corner and the P_{DC} value due to another corner. Typically, and as expected, the designs have higher phase noises and higher power consumption when simulated in its corners.

A more complete analysis of how many designs failed to meet the performances after corner simulation was also performed. The results are shown in Fig. 4.24. It shows that the most affected performances by corners are the oscillation frequency and the phase noise nearer the carrier (<100 kHz offset), which correspond to the Flicker (or $1/f$) noise region [10]. This increase in the Flicker noise is most affected

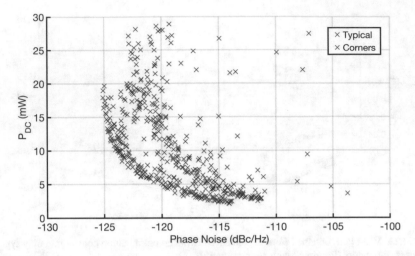

Fig. 4.23 PN vs. PDC projection of the VCO typical POF (from Fig. 4.22) compared with the same designs simulated with its corners (worst corner performance depicted) (Reprinted with permission [9])

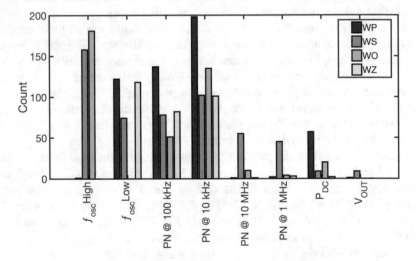

Fig. 4.24 Histogram showing the number of designs (Y-axis) that fail each constraint at each considered corner (Reprinted with permission [9])

due to the WP corner; this seems plausible, since this type of noise is directly related to the direct current flowing through the devices, and, therefore, its power consumption.

Furthermore, in Table 4.15, five points from the POF are shown in more detail, with their typical and worst corner performances. It can be noticed that some of the performances are over-estimated in a pre-corner analysis (e.g., PN), while others are under-estimated (e.g., P_{DC}).

Table 4.15 Desired performances and specifications for the VCO optimization. Comparison between the typical corner and worst corner performances for five different designs

Performances	Specifications	Design 1		Design 2		Design 3		Design 4		Design 5	
		Typical	WC[a]	Typical	WC[a]	Typical	WC[a]	Typical	WC[a]	Typical	WC[a]
f_{osc} (Vtune= 0 V)	>2.55 GHz	2.551	2.429	2.553	2.406	2.666	2.511	2.746	2.586	2.556	2.420
f_{osc} (Vtune= 2.5 V)	<2.45 GHz	2.325	2.441	2.135	2.262	2.268	2.397	2.372	2.512	2.284	2.415
PN @ 1 MHz	<−113 dBc/Hz	−125.1	−122.8	−123.5	−120.0	−119.2	−115.8	−117.4	−114.2	−114.8	−103.7
P_{DC}	<20 mW	17.9	25.9	11.1	14.8	5.1	5.8	6.7	7.4	3.1	3.6
V_{OUT}	>0.15 V	1.83	1.51	1.39	0.98	1.09	0.93	0.75	0.50	0.6	0.16

[a]Worst corner performance: in red the ones that fail to meet the constraint and in green the ones that meet the constraint

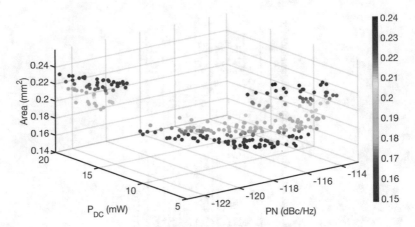

Fig. 4.25 VCO POF obtained from the enhanced two-step optimization considering all corners (worst corner performances depicted) (Reprinted with permission [9])

This means that although this approach is already a step forward towards robustness, due to the modeling technique used for inductors, which already incorporates detailed inductor parasitics, it is still not enough. Consequently, in order to increase the solution robustness even more, the device corners must be considered during optimization.

Therefore, an optimization was performed in the same conditions presented in Table 4.14, but also considering the above-mentioned technology process corners. The new optimization was performed simulating all corners for each individual VCO and ensuring constraints on all corners. The obtained POF can be seen in Fig. 4.25.

The optimization took approximately 10 h in an Intel Core i7-3770 @ 3.4 GHz workstation with 32 Gb of RAM, which is twice the computation effort than the previous optimization presented in Fig. 4.22, but the obtained POF is much more reliable. It can be concluded that relatively poorer phase noises are achieved with similar power consumptions; however, the biggest price to pay is area, as it seems that the designs usually use larger device sizes in order to comply with the constraints in all corners. In order to study this issue, the device sizes of the VCOs from the POF considering corners were compared with the device sizes of the VCOs from the POF without corners. The comparisons can be seen in Fig. 4.26. It is possible to observe that almost every device needs larger area in order to comply with constraints when the corner analysis is considered. This is particularly noticeable in the area of the capacitors and varactors. In order to comply with the tuning range specifications in all extreme corners, the varactors need larger area. Consequently, larger varactors have higher capacitance value, and, therefore, in order to maintain the same f_{osc}, the inductance has to be smaller (see Eq. (4.2)), and this is why the inductors used in the POF considering corners use inductors with only two turns (which typically have lower inductance values than inductors with more turns).

Afterwards, a more detailed analysis of one of the VCOs is also performed (the chosen VCO has the lowest phase noise from the POF shown in Fig. 4.25). The design was simulated in its typical and extreme corners. The performance variations of the circuit among all these corners can be observed in Fig. 4.27. This design complies with all design specifications at all corners, and is, therefore, very robust.

It can be concluded that by using this so-called corner-aware optimization approach, more reliable and robust designs can be achieved compared to the optimization that solely considers the typical device models, bringing the designer closer to a first-pass fabrication success.

4.4 Mixer Design using a Systematic Circuit Design Methodology

This section depicts the optimization-based design of a Gilbert cell mixer just for the sake of illustration and to demonstrate the design of each circuit of the RF front-end.

4.4.1 Experimental Results: Multi-Objective Optimizations

The mixer topology considered is shown in Fig. 4.28a. It is a Gilbert cell mixer, intended to down-convert the RF frequency from 2.46 GHz to 40 MHz. The testbench used for the mixer optimization is shown in Fig. 4.28b where an ideal RF signal at 2.46 GHz was set as well as an ideal local oscillator (LO) frequency at 2.5 GHz. The mixer will operate with a supply voltage of V_{DD}=2.5 V.

The mixer performances are the most difficult to calculate/simulate from all the presented circuits. This is due to the fact that the mixer receives two input signals at two different frequencies, which makes the analysis setup more complex. Firstly, a PSS analysis must be performed in order to calculate the steady-state of the circuit. Afterwards, for the conversion gain (CG), a Periodic-AC (PAC) analysis is performed. The CG will depend on the IF frequency at which the mixer will operate. Therefore, this PAC analysis can be performed by sweeping the RF frequency range in order to inspect how the CG changes. For the mixer NF calculation, a noise analysis can be performed. This NF also changes with the IF frequency, and, therefore, a sweep can also be performed in order to check the NF value for different IF frequencies. For the port-to-port isolation, a periodic transfer function analysis (PXF) is performed which is able to detect the influence of the signals generated by the LO and RF source in the IF source and vice-versa. By doing so, it is possible to consider the influence of the unwanted signals in each port. Last but not the least, the mixer IIP_3 is also very problematic and difficult to properly calculate.

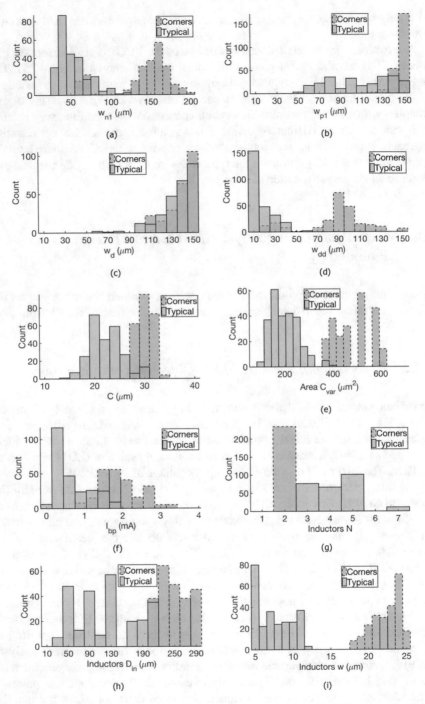

Fig. 4.26 VCO device sizes from the POF considering corners versus the POF with only typical models (Reprinted with permission [9])

Fig. 4.26 (continued)

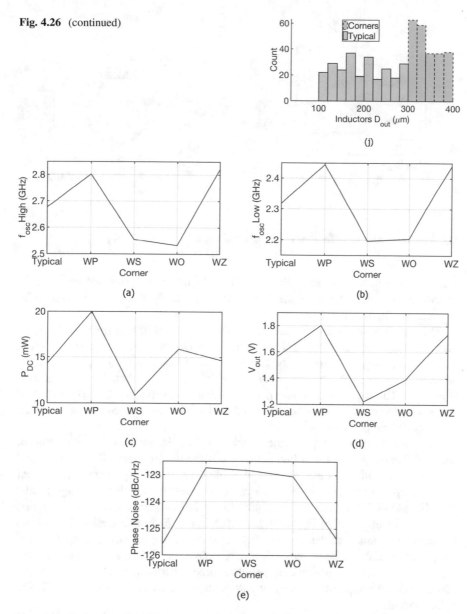

Fig. 4.27 VCO performance variations for one design of the obtained POF, both the typical and the corner models (Reprinted with permission [9])

The mixer has two input frequencies (RF and LO), which makes the PSS+PAC method used in the LNAs invalid, because the PSS does not support so many input frequencies. Therefore, in order to calculate the mixer IIP_3, modern simulators, such as SpectreRF [11], allow Quasi-PSS and Quasi-PAC analysis which together

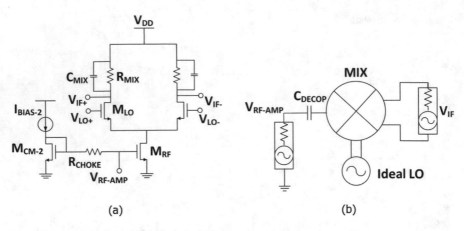

Fig. 4.28 (a) Schematic of the mixer and (b) its testbench

enable the calculation of the IIP_3. Apart from the different performed analysis, the method to efficiently calculate the IIP_3 is the same as for the LNAs, used in order to avoid any input power sweep.

The circuit design variables are shown in Table 4.16. In this specific topology, there are 12 initial design variables: the n-type MOS transistors sizes (w_{LO}, w_{RF}, w_{CM-2}, l_{LO}, l_{RF}, and l_{CM-2}), the resistor widths ($R_{W_{CHOKE}}$, $R_{W_{MIX}}$) and lengths ($R_{l_{CHOKE}}$ and $R_{l_{MIX}}$), the capacitor value (C_{MIX}), and also, the bias current (I_{BIAS-2}). The capacitance C_{DECOP} is a decoupling capacitance, and not part of the design; therefore, it is maintained fixed at 10pF because it does not influence the mixer performances.

The optimization was performed with 256 individuals and 250 generations, and had three objectives: minimization of power consumption, NF and area, and the desired specification can be seen in Table 4.17. In this optimization, in order to speed up the process, the CG is only calculated at two IF frequency points: 10 MHz and 40 MHz. The same applies to the NF calculation. The optimization results can be observed in Fig. 4.29. However, in this first optimization the process variability was not considered. Therefore, in order to inspect how damaging could the process variability be to the designs available in the POF, all the mixer designs were re-simulated considering their extreme corner performances. Like in the VCO, considered in the previous section, the following corners were considered: WP, WS, WO, and WZ.

Similarly to the VCO, it was found that from the 256 mixer designs obtained in the POF from Fig. 4.29, none of them complied with constraints after being evaluated in its performance corner conditions. Therefore, in practical terms there would be no actual POF because all the designs were invalid. In Fig. 4.30a it is possible to observe the comparison between the mixer typical POF (from Fig. 4.29) compared with the same designs simulated with its corners (worst corner performance depicted). Since in 3D it is difficult to observe the shifts in the

Table 4.16 Design variables
for the mixer optimization

Variables	Min	Max	Grid
$w_{LO,RF,CM-2}$ (μm)	10	200	10
$l_{LO,RF,CM-2}$	Fixed @ 0.35 μm		
I_{BIAS-2} (mA)	0.1	1.5	0.1
$R_{W_{CHOKE,MIX}}$ (μm)[a,b]	1	3	1
$R_{l_{CHOKE,MIX}}$ (μm)[a,b]	3	90	1
C_{MIX} (pF)	0.3	3	0.3
C_{DECOP}	Fixed @ 10pF		

[a]Three series resistances are used for R_{MIX} in order
to increase the ranges. The min and max values shown
are for each resistor
[b]Four series resistances are used for R_{CHOKE} in
order to increase the ranges. The min and max values
shown are for each resistor

Table 4.17 Design
specifications for the mixer
optimization

Mixer performance	Mixer specifications
CG @ 10 MHz	>5 dB
CG @ 40 MHz	>5 dB
P_{DC}	Minimize
P_{DC}	<25 mW
NF @ 10 MHz	<20 dB
NF @ 40 MHz	Minimize
NF @ 40 MHz	<20 dB
IIP_3	>5 dBm
Port-to-Port Isolation	<-30 dB
Area (μm²)[a]	–

[a] The area of the mixers is not considered for
the optimization because this circuit does not have
any inductor. Therefore, its area is relatively small
when compared to the VCO or the LNA

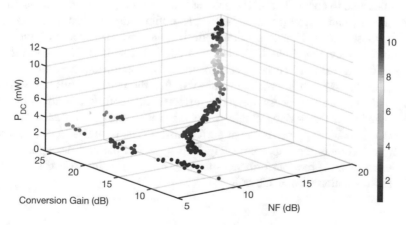

Fig. 4.29 Mixer POF obtained considering only nominal corner performances

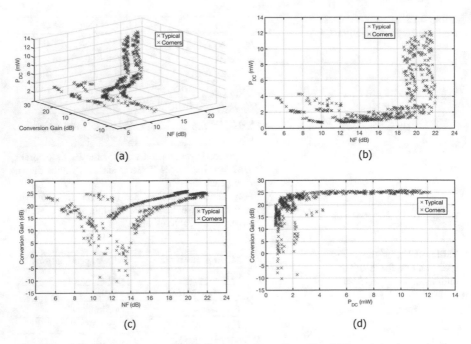

Fig. 4.30 Comparison between the mixer typical POF (from Fig. 4.29) and the same designs simulated with its corners (worst corner performance depicted). (**a**) 3D comparison. (**b**) 2D projection. NF vs. P_{DC}. (**c**) 2D projection. NF vs. CG. (**d**) 2D projection. P_{DC} vs. CG

performances, 2D projections were performed in order to ease the POF comparison (see Fig. 4.30b–d). It should be taken into account that the corner performances depicted are the worst possible for each corner of the design.

Typically, and as expected, the designs have higher NF, higher power consumption, and less CG when simulated in its corners.

It is possible to conclude that the process variability highly degrades the circuit performances, and, therefore, a corner-aware optimization is desired in order to consider these effects. Hence, another optimization was performed with the same conditions as shown in Table 4.17, also with 256 individuals and 250 generations, but considering extreme corner performances. The optimization results can be observed in Fig. 4.31. It is possible to observe that the feasible objective space is highly reduced due to process variability effects. Although the feasible objective space is reduced, the designs obtained are far more robust to the ones obtained in the POF considering only nominal performances. The optimization considering only typical corners took approximately 19 h of CPU time and the optimization considering worst corner performances took 33 h, using an Intel Core i7-3770 @ 3.4 GHz workstation with 32 Gb of RAM.

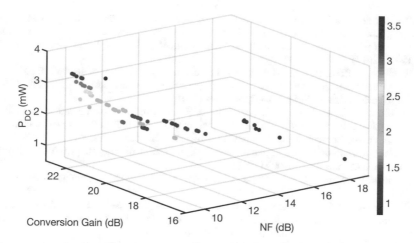

Fig. 4.31 Mixer POF obtained considering worst corner performances

4.5 Summary

In this chapter the influence of using different inductor modeling techniques was studied. This was performed for an LNA case study. Both a physical and a surrogate model were integrated in optimization-based approaches in order to design the inductors presented in the adopted LNA topology. It was shown that it is imperative to use highly accurate inductor models and that surrogate models present a better solution. Moreover, a hierarchical bottom-up design methodology was proposed in order to design an LNA. A study was performed between hierarchical and non-hierarchical optimization-based methodologies. It was seen that hierarchical optimization-based methodologies are able to achieve superior results.

At the end of the chapter, it was illustrated how process variability poses a significant problem and the impact they can have within optimization-based methodologies.

References

1. H. Zhang, E. Sanchez-Sinencio, Linearization techniques for CMOS low noise amplifiers: a tutorial. IEEE Trans. Circuits Syst. I Regul. Pap. **58**, 22–36 (2011)
2. F. Passos, E. Roca, R. Castro-López, F. Fernández, A two-step surrogate modeling strategy for single-objective and multi-objective optimization of radiofrequency circuits. Soft Comput. **23**, 4911–4925 (2019)
3. F. Passos, E. Roca, R. Castro-López, F. Fernández, Radio-frequency inductor synthesis using evolutionary computation and gaussian-process surrogate modeling. Appl. Soft Comput. **60**, 495–507 (2017)

4. R. Gonzalez-Echevarria, E. Roca, R. Castro-López, F.V. Fernández, J. Sieiro, J.M. López-Villegas, N. Vidal, An automated design methodology of RF circuits by using pareto-optimal fronts of EM-simulated inductors. IEEE Trans. Comput. Aided Des. Integr. Circuits Syst. **36**, 15–26 (2017)
5. F. Passos, E. Roca, R. Castro-López, F.V. Fernández, A comparison of automated RF circuit design methodologies: Online versus offline passive component design. IEEE Trans. Very Large Scale Integr. VLSI Syst. **26**, 2386–2394 (2018)
6. M. Velasco-Jiménez, R. Castro-López, E. Roca, F.V. Fernández, Implementation issues in the hierarchical composition of performance models of analog circuits, in Design, Automation and Test in Europe Conference and Exhibition (2014), pp. 12:1–12:6
7. D. Ham, A. Hajimiri, Concepts and methods in optimization of integrated LC VCOs. IEEE J. Solid State Circuits **36**, 896–909 (2001)
8. L. Fanori, P. Andreani, Class-D CMOS oscillators. IEEE J. Solid State Circuits **48**, 3105–3119 (2013)
9. F. Passos, R. Martins, N. Lourenço, E. Roca, R. Póvoa, A. Canelas, R. Castro-López, N. Horta, F. Fernández, Enhanced systematic design of a voltage controlled oscillator using a two-step optimization methodology. Integration **63**, 351–361 (2018)
10. C. Samori, Understanding phase noise in LC VCOs: a key problem in RF integrated circuits. IEEE Solid-State Circuits Mag. **8**(4), 81–91 (2016)
11. SpectreRF. https://www.cadence.com/content/cadence-www/global/en_US/home/tools/custom-ic-analog-rf-design/circuit-simulation/spectre-rf-option.html. Accessed: 28-01-2020

Chapter 5
Systematic Circuit Design Methodologies with Layout Considerations

In Chap. 4, it was demonstrated that the modeling of passive components is very important for an accurate optimization-based design strategy. Furthermore, the offline strategy to design such passive components has demonstrated to provide superior results and more efficiently. Also, the device corner performances have shown to have a high impact on the performances of the circuits, and, therefore, must be considered from a first design stage. In this Section, another important issue that should be considered in systematic circuit design methodologies is addressed: the layout parasitics. These layout parasitic effects, that appear in circuits (especially RF circuits), are in many cases devastating, and most times cause the circuit performances to degrade immeasurably. Hence, the automated sizing procedures which do not consider these effects are usually not enough to achieve robust circuit designs. Therefore, it is clear that the layout parasitic effects must also be taken into account during the design stage.

In this chapter a layout-aware optimization-based design methodology is developed to systematically design RF basic blocks tackling the layout parasitics problem in a comprehensive way, where the routing parasitics and the parasitics from all critical passive devices (i.e., inductors) are considered. In Sect. 5.1, the framework on which the methodology is implemented is presented. Afterwards, the methodology is explained in detail in Sect. 5.2. In Sect. 5.3, the methodology will be applied to the design of each block forming an RF front-end: VCO, LNA, and mixer, considering all layout parasitic effects. Since the extreme corner performances are also very important during an initial design stage, in this chapter a methodology is developed where these effects are considered in parallel with the layout parasitics. Therefore, in Sect. 5.4, a VCO will be designed considering the layout parasitics and the extreme corner performances in order to achieve highly robust designs, intended for a first-pass fabrication success.

© Springer Nature Switzerland AG 2020
F. Passos et al., *Automated Hierarchical Synthesis of Radio-Frequency Integrated Circuits and Systems*, https://doi.org/10.1007/978-3-030-47247-4_5

5.1 AIDA Framework: AIDA-L

The layout-aware design methodology has been implemented in the AIDA framework [1]. AIDA is a standalone tool which allows a fully automated design from circuit-level specifications to physical layout description. AIDA presents itself as two different modules that communicate with each other. The first module, AIDA-C, assists the analog designer in the sizing of analog circuits. The second module, AIDA-L, allows the user to automatically create layouts during the optimization loop. When combined, these two modules, AIDA-C and AIDA-L, give the user the capability of performing optimization-based synthesis of a given circuit with layout considerations. In this Chapter, the AIDA-L capabilities are used and, therefore, its complete architecture is shown in Fig. 5.1.

The AIDA-L tool was originally developed to automatically perform the layout of circuits operating in baseband, such as operational amplifiers. However, the layout issues in baseband are far less critical than in RF. Therefore, the AIDA-L paradigm had been changed in order to support RF circuits. In baseband, while performing the layout of the circuit, the designer is usually interested in e.g., compacting the complete layout in order to obtain the best area minimization, taking into account electromigration issues, etc. However, in RF circuits some other issues have to be taken into account. In RF, while the area is also important, the designer has to take into account the effect of several other issues, such as symmetry of RF signal paths. Another issue that must be tackled in RF is related to the routing

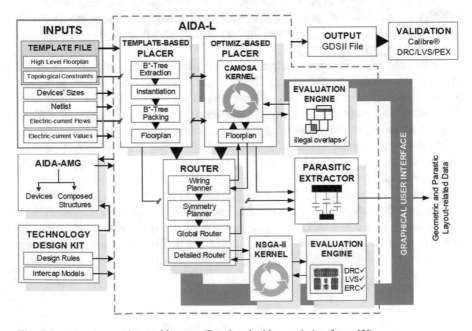

Fig. 5.1 AIDA-L complete architecture (Reprinted with permission from [2])

width. While in baseband the routing width can be minimized as much as possible, while complying with DRC rules, electromigration, etc., in RF, especially in RF signal paths, the routing must be wider and performed in less resistive metals (upper metals) in order not to degrade the signal integrity. Therefore, in order to support the automatic layout-aware design of RF circuits, some changes and improvements were performed over this framework. These improvements are described in the next Section, while describing the developed methodology.

5.2   RF-Specific Layout-Aware Circuit Design Methodology

The general description of the methodology is illustrated in Fig. 5.2. The layout-aware methodology follows the two-step enhanced bottom-up strategy presented in Chap. 4. The methodology uses the previously presented surrogate model for the inductor POF generation and performs a layout-aware sizing optimization, the latter encompassing several other complex tasks. Following a bottom-up approach, the two distinct phases communicate through the generated integrated inductor POFs. Since the surrogate modeling strategy was explained in Chap. 3, and the two-step design methodology was explained in Chap. 4, in this Chapter their explanation is skipped. Therefore, only the complex tasks performed in the automatic layout-aware optimization are described.

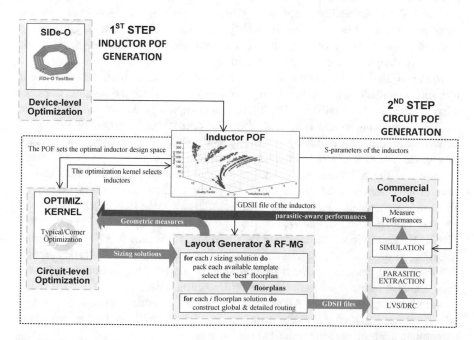

Fig. 5.2 Detailed description of the proposed two-step RF-specific layout-aware circuit sizing loop

The circuit optimization is also carried out by the NSGA-II optimization algorithm, that selects different sizing solutions, each one with a new set of design variables (i.e., transistors widths, lengths or number of fingers, etc.). As previously mentioned, the inductor POF is generated *a priori*. The layout of these obtained inductors is automatically generated and saved as GDSII files. While performing an optimization, the inductors are imported as GDSII files, and the remaining device modules (e.g., transistors, resistors, etc.) are generated using an internal RF module generator (MG) available in AIDA-L.

The methodology uses user-defined templates in order to define the floorplan. Since any geometrical requirement can be imposed to the layout as a constraint or an objective for the optimization problem, the floorplan that suits most of those requirements for each sizing is passed to the router. Here, a wiring topology, global routing, and detailed routing solutions with wiring symmetry are devised for each floorplan.

5.2.1 RF Template-Based Placer

The AIDA-L RF template-based placer is built over the template-based approach for analog and mixed-signal IC blocks introduced in [3], where the designer is responsible for providing a high-level floorplan description, and then, the instantiation is automatically performed given any set of device sizes. Similarly, the RF MG has the ability to construct structures such as double-poly capacitors, varactors, RF metal-insulator-metal (MIM) capacitors, RF MOSFETs, guard rings shielding around the entire circuit in order to minimize signal couplings, etc.. This RF MG is completely technology-independent.

In the AIDA-L RF template-based placer, any device can be imported as a GDSII file, such as inductors, PADs, or even a part of the circuit which the designer does not wish to optimize (e.g., buffers). The high-level floorplan guidelines are provided in a XML form, using simple structures where the user is able to define the symmetry between structures, the matching, proximity, and spatial relations that he/she wishes to impose. An example of a template description for a generic VCO schematic is shown in Fig. 5.3. As it is shown, the template can be partitioned into several smaller templates in order to add hierarchy to the template. Every time additional constructs are needed, e.g., N-type/P-type guard ring around PMOS/NMOS devices and inductor shielding around the complete block, this hierarchy should be updated. The XML description of the template shown in Fig. 5.3 is shown in Fig. 5.4. Further information on this XML description can be found in [2]. This XML description defines a set of guidelines, which are then used to generate the circuit floorplan for each tentative sizing solution during the optimization flow.

Another improvement to AIDA-L is the ability to define *no-crossing* boxes, which are used to keep consistent distances in all four directions during automatic placement, attempting to minimize not only the magnetic coupling between devices, but also to the rest of the circuit structures (e.g., PADs). This is especially useful

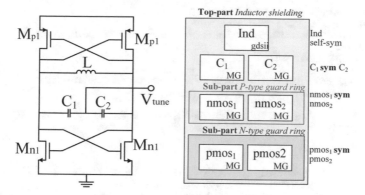

Fig. 5.3 General VCO schematic (without biasing circuitry) and a graphical representation of the template hierarchy used for the layout generation. Devices are generated using the custom module generator or imported as a GDSII file (gdsii)

for circuits using integrated inductors where the coupling between them is a problematic issue. These no-crossing boxes are illustrated in Fig. 5.5.

5.2.2 RF Fully-Automatic Router

While the placement is controlled by the circuit designer in a high-level fashion using the technology- and specification-independent templates, all routing setup and execution is kept independent from placement. That is, as solutions vary in a multitude of different device sizes/shapes and performances, regardless of the template and subsequent placement obtained, the router presents a design-rule- and layout-versus-schematic-correct layout solution, and, therefore, no additional effort is needed from the designer. To accomplish this, a three-step router is adopted.

5.2.2.1 Wiring Planner

The first block parses the netlist of the RF block under optimization, and, together with the obtained floorplan, derives the wiring topology that connects all the terminals and provides the optimal terminal-to-terminal connectivity. The problem is formulated as: given a set of t terminals $\{T1, T2,\ldots, T_t\}$ construct the tree that is strongly connected (i.e., all terminals are interconnected) and that minimizes the total wiring area of the tree given by

$$\sum_{i=1}^{t}\sum_{j=1}^{t} l_{i,j} \times w_{i,j}, i \neq j \tag{5.1}$$

```xml
<?xml version="1.0" encoding="ISO-8859-1"?>
<!DOCTYPE Template SYSTEM "template4.dtd">
<Template name="VCO_example" electromigration="no" pads="no" powernets="no">
    <CellList>
        <-- INDUCTOR -->

        <Cell name="INDUCTOR" symGroupId="1" symCellId="-1" rotate="MY">
            <Box x="0" y="3000" />
            <SIDeO idx="l0_idx"
                path="$PDK_DIR/ams_c35/models/SIDeO/SPIRAL_SYM_OCT_GR@2.4GHZ/"></SIDeO>
        </Cell>

        <-- CAP -->

        <Cell name="CAP1" symGroupId="1" symCellId="2" rotate="RCCLK_90">
            <Box x="0" y="2000" w="1000" h="1000" />
            <CPOLY width="RF_cap" length="RF_cap"></CPOLY>
        </Cell>

        <Cell name="CAP2" symGroupId="1" symCellId="2" rotate="RCCLK_90">
            <Box x="1000" y="2000" w="1000" h="1000" />
            <CPOLY width="RF_cap" length="RF_cap"></CPOLY>
        </Cell>
    </Cell>

        <-- NMOS -->

    <Cell name="NMOS1" symGroupId="1" symCellId="3">
        <Box x="0" y="1000" w="1000" h="1000" />
        <MOSFET type="N" width="10e-6*NG_NM2" length="0.35e-6" nf="NG_NM2" />
    </Cell>

    <Cell name="NMOS2" symGroupId="1" symCellId="3">
        <Box x="1000" y="1000" w="1000" h="1000" />
        <MOSFET type="N" width="10e-6*NG_NM2" length="0.35e-6" nf="NG_NM2" />
    </Cell>

        <-- PMOS -->

    <Cell name="PMOS1" symGroupId="1" symCellId="4" rotate="RCCLK_90">
        <Box x="0" y="0" w="1000" h="1000" />
        <MOSFET type="P" width="10e-6*NG_CM" length="0.35e-6" nf="NG_CM" />
    </Cell>

    <Cell name="PMOS2" symGroupId="1" symCellId="4" rotate="RCCLK_90">
        <Box x="1000" y="0" w="1000" h="1000" />
        <MOSFET type="P" width="10e-6*NG_PM1" length="0.35e-6" nf="NG_PM1" />
    </Cell>
    </CellList>
</Template>
```

Fig. 5.4 XML template file which describes the template shown in Fig. 5.3

where $l_{i,j}$ is the rectilinear distance between two terminals and $w_{i,j}$ is the wire
width. This problem is solved with the method proposed in [4]. When all wiring
topologies are obtained, the symmetry information of the high-level floorplan
guidelines is used to identify symmetries between terminal-to-terminal instances.
Each terminal-to-terminal connection from each wiring topology (i.e., each net) is
tested with all other connections from the same and other wiring topologies.

Fig. 5.5 Illustrating the use of no-crossing boxes in the template hierarchy

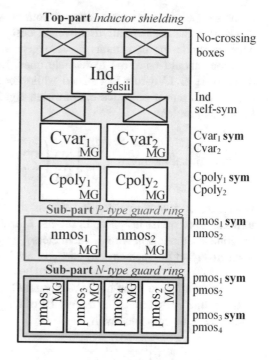

5.2.2.2 Single-Net Global Router

Each terminal of each cell may contain multiple electrically equivalent locations where the connection can be made. Therefore, in the second step of the router, each terminal-to-terminal connection is transformed into a rectilinear path where the ports that minimize the wire area, while implementing the detected symmetries, are selected from the multiple available ports of their corresponding terminals. To achieve this, first, a multilayer rectilinear grid for the problem is constructed, which consists on a directed graph considering three-dimensional vertices and obstacles. Then, to select the best ports to be connected and the corresponding path, a variation of the A* path-finding algorithm [4] operating over the sparse non-uniform multilayer grid is taken.

5.2.2.3 Multi-Net Detailed Router

In the last step, the previously determined terminal-to-terminal connections, global routing solution, and symmetry information are used as the starting point for an optimization process. This optimization-based routing follows the principles introduced in [3], where an evolutionary algorithm performs slight structural and layer changes in the physical representation of a population of different and independent detailed routing solutions. This allows optimizing all wires of all nets simultaneously. The different routing solutions are evaluated by built-in algorithms that are basically lightweight design-rule- and LVS-check procedures, which provide the constraints at each generation of the optimization problem. Solutions should converge to feasible ones, i.e., feasibility is attained when no design-rule or LVS-rule is violated while keeping the minimum wire length as objective. A single-objective optimization algorithm is used in order to solve the following problem:

$$\text{find } x_d \text{ that minimize } \sum_{i=1}^{N_{net}} \left[\sum_{j=1}^{n_i} \left(\sum_{k=1}^{k_j} \left(l_{i,j,k} \times w_{i,j,k} \times z_{i,j,k}\right) \right) \right] \qquad (5.2)$$

where, x_d is a routing solution, N_{net} is the number of nets, n_i the number of terminal-to-terminal instances in the net i, and k_j stands for the fixed number of rectilinear segments in wire j. Parameters $l_{i,j,k}$ and $w_{i,j,k}$ are the length and width, respectively, of the segment k of wire j of net i, and $z_{i,j,k}$ the pre-defined "cost" of a given metal layer (preferred layers should be associated with the lower costs). Since the routing solution must not have any DRC or LVS errors, these are the routing optimization constraints. The dimension of the design variables space for a routing solution can be computed with

$$x_d = \sum_{i=1}^{N_{net}} \left[\sum_{j=1}^{n_i} \left(\sum_{k=1}^{k_j} \left(l_{i,j,k} + layer_{i,j,k}\right) + S_{rc} + S_{ink} \right) \right] \qquad (5.3)$$

with $layer_{i,j,k}$ being the index of a different available conductor, and, S_{rc} and S_{ink} the different electrically equivalent locations available within the start or end terminals, respectively, of a wire to establish the connection.

5.3 Optimization-Based Synthesis with Layout Considerations

5.3.1 VCO Design Using a Bottom-Up Systematic Layout-Aware Methodology

In this Section, the results obtained using the proposed methodology for the design of a cross-coupled double-differential VCO are presented. The VCO is the same used in Sect. 4.3, however, in this case the VCO is connected to output buffers to ease the experimental characterization, accounting for future fabrication (see Fig. 5.6). The adopted technology is the same $0.35\,\mu$m-CMOS technology used in previous experiments. However, the methodology itself is completely independent of the adopted technology.

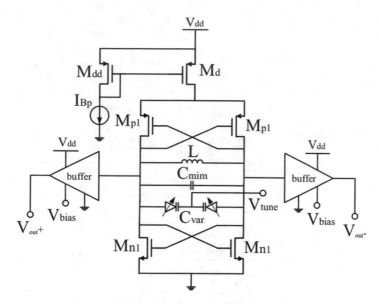

Fig. 5.6 Cross-coupled double-differential VCO schematic with output buffers. (Reprinted with permission [5])

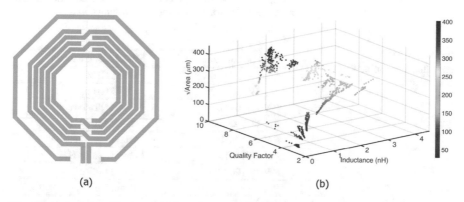

(a)   (b)

Fig. 5.7 Inductor topology and POF used in the VCO layout-aware optimization. (**a**) Instance of the octagonal symmetric inductor with guard ring. (**b**) POF of 1000 points with the trade-off $\sqrt{\text{Area}}$ vs. Quality factor vs. Inductance of the octagonal symmetric inductor with guard ring

The inductor topology chosen for this circuit is the octagonal symmetric inductor with guard ring (shown in Fig. 5.7a), and the design variables of the inductor used to generate the POF are shown in Table 5.1. A multi-objective optimization was performed with 1000 individuals and 80 generations, resulting in the POF shown in Fig. 5.7b.

The circuit design variables and the optimization objectives and constraints are shown in Tables 5.2 and 5.3, respectively. The design variables ranges are selected according to the minimum and maximum available by the foundry. Also, the grid of

Table 5.1 Inductor design variables

Parameter	Minimum	Grid	Maximum
N	1	1	8
D_{in} (μm)	10	1	300
w (μm)	5	0.05	25
s (μm)	2.5	–	2.5

Table 5.2 Design variables of the VCO

Variables	Minimum	Maximum	Grid
w_{n1} (μm)	10	200	10
$w_{p1,d,dd}$ (μm)	10	150	10
$l_{n1,p1,d,dd}$ (μm)	Fixed @ 0.35 μm		
I_{bp} (mA)	0.1	1.5	0.1
w_{Cvar} (μm)	Fixed @ 6.6 μm		
l_{Cvar} (μm)	Fixed @ 0.65 μm		
Inductors	Selected from the POF		
Row_{Cvar}, Col_{Cvar}[a]	4	12	1
C (μm)[b]	9	29	1

[a]Row_{Cvar} and Col_{Cvar} are the number of fingers per row and column of the varactors
[b]In this Chapter, six parallel capacitors are used. The min and max values are for each of the six capacitors used

Table 5.3 Design specifications for the VCO optimization

VCO performance	VCO specifications
f_{osc} ($V_{tune} = 0$ V)[a]	>2.55 GHz
f_{osc} ($V_{tune} = 2.5$ V)[a]	<2.45 GHz
PN @ 10 kHz	<−65 (dBc/Hz)
PN @ 100 kHz	<−92 (dBc/Hz)
PN @ 1 MHz	Minimize
PN @ 1 MHz	<−113 (dBc/Hz)
PN @ 100 MHz	<−134 (dBc/Hz)
P_{DC}	Minimize
P_{DC}[b]	<40 mW
V_{OUT}	>0.15 V
Area[c]	Minimize

[a]By applying these constraints we ensure that the VCO covers the target oscillation frequency of 2.5 GHz, ensuring at the same time an acceptable tuning range
[b]The constraint on P_{DC} was increased when compared to the optimizations in Chap. 4 due to the usage of the buffers
[c]The area of each VCO is approximated as the sum of all device sizes

the variables is selected according to the foundry information. For the capacitances, a MIM bank was used, consisting of 6 parallel capacitors, in order to provide a wide capacitance range. The circuit is intended to operate at $V_{dd} = 2.5$ V.

5.3.1.1 Sizing Optimizations

To first motivate the need for a layout-aware flow, ten runs of the sizing optimization were performed. In these optimizations, only the sizing of the devices as well as the EM-characterized inductors were considered, but disregarding full parasitic extraction. It is important to note that the inductor POF provides an accurate parasitic estimation and the foundry models of the devices already contemplate an estimation of the layout parasitics of the devices; therefore, the pre- to post-layout gap is already shortened. Each optimization was performed with a population of 256 individuals and 200 generations. Since no layout is available yet, the minimization of the sum of the area of the devices (including buffers and PADs) is adopted instead of the real placement area. A summary of the results of these 10 runs is presented on columns three to eight in Table 5.4, and the POF obtained in one run, run_{VCOS2}, is illustrated on Fig. 5.8.

Table 5.4 Minimum and maximum objective values of the POFs obtained in ten sizing optimizations. Post-layout simulation of the previous obtained POFs

| Run | Feasible solutions | Minimum and maximum objective values of the POF | | | | | | Post-layout[a] | | |
| | | Area (mm^2) | | PN@1 MHz (dBc/Hz) | | P_{DC} (mW) | | | | |
		Min	Max	Min	Max	Min	Max	Feasible	Fail[b]	n/c[c]
run_{VCOS1}	256	0.1736	0.3809	−127.84	−115.41	12.771	37.524	6	23	228
run_{VCOS2}	256	0.1730	0.3908	−128.10	−114.76	12.598	39.835	6	27	223
run_{VCOS3}	256	0.1700	0.3909	−127.86	−114.90	12.766	39.490	5	29	222
run_{VCOS4}	256	0.1741	0.3926	−128.06	−114.96	12.720	36.903	6	19	231
run_{VCOS5}	256	0.1722	0.3819	−128.20	−114.88	12.674	39.973	3	36	217
run_{VCOS6}	256	0.1725	0.3518	−127.49	−114.75	12.496	39.771	17	30	209
run_{VCOS7}	256	0.1744	0.3876	−128.36	−114.91	12.611	39.973	6	41	209
run_{VCOS8}	256	0.1728	0.3826	−128.22	−114.81	12.545	39.952	9	25	222
run_{VCOS9}	256	0.1733	0.3803	−127.72	−115.02	12.630	39.566	8	35	213
run_{VCOS10}	256	0.1739	0.3756	−127.70	−114.94	12.927	37.182	11	22	223

[a]Layouts automatically generated using the methodology of Sect. 5.2 and extracted with Calibre xRC
[b]Solutions that failed at least one constraint
[c]Steady-state analysis not converged

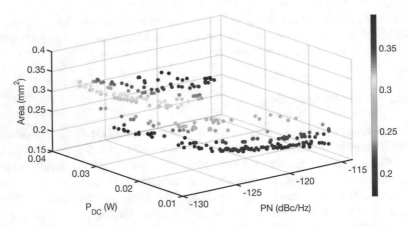

Fig. 5.8 VCO Sizing optimization: Pareto front showing the trade-offs area vs. PN@1 MHz vs. P_{DC} of run_{VCOS2} (Reprinted with permission from [5])

5.3.1.2 Post-Layout Simulations

After obtaining the sizing solutions, the placement template hierarchy of Fig. 5.9 was defined, taking into account several aspects important for RF circuits: symmetric signal paths, routing minimization, mismatch between devices, and also, all the following structures were considered: guard rings, shielding, dummies, PAD placement, etc.

The layouts of the designs from the ten previous runs (i.e., 2560 designs) were automatically generated with the procedures proposed in Sect. 5.2. The parasitics of each automatically generated layout were extracted with Mentor Graphics Calibre [6] and the full post-layout performances measured. The post-layout verification results are shown in the last three columns of Table 5.4. From the original 2560 designs, in approximately 86% of the points the steady-state analysis failed convergence, while 11% actually got simulated but failed to meet at least one performance constraint post-layout, and only a small set verify all performance constraints. The high number of solutions that fail to achieve steady-state convergence may be due to two different issues. On the one hand, the presence of layout parasitic devices, such as parasitic resistances and capacitors, degrade the circuit performances and may prevent the oscillator from having sufficient g_m in order to oscillate. On the other hand, while setting the steady-state analysis, the designer must specify a guess oscillation frequency that should be relatively near from the actual oscillation frequency. Therefore, when performing post-layout simulations, it may happen that these parasitic devices, swerve the f_{osc} in such a severe way that the steady-state is no longer achieved. But if this is the case, it means that the real f_{osc} of the design is very different from the desired one, and, therefore, it is not a valid design for our purposes. Tweaking the guessed oscillation frequency to make the simulator converge to a solution is not interesting for optimization purposes, because it would take additional computation time without any benefit (since the design has its f_{osc}

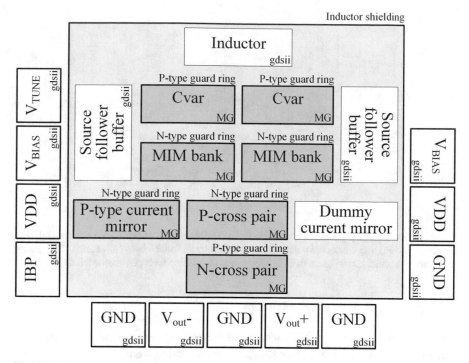

Fig. 5.9 Placement template hierarchy for the VCO. Shaded boxes refer to sub-partitions. Devices, blocks, or sub-partitions are either generated using the custom module generator or imported as a GSDII file (gdsii). (Reprinted with permission from [5])

far from the desired one). Therefore, in the adopted approach, when any of the previous problems emerge, the sizing solution is treated as a design failure and the individual VCO is discarded.

Particularly for run_{VCOS2}, the post-layout front is illustrated in Fig. 5.10 and in Table 5.5 pre- and post-layout performances of three sample points from this front are presented. In Fig. 5.11, the layout of point S_2 (from Table 5.5) is illustrated.

As it is clearly observable, the disregard for a full parasitic extraction leads to some of the performances being over-estimated in a pre-layout stage (e.g., P_{DC}), others being mostly under-estimated (e.g., PN, V_{OUT}), and the oscillation frequency always deviating from pre-layout values. In the traditional flow, the layout of each design failing post-layout simulation would be re-iterated manually until, if even achievable, the impact of layout parasitics is negligible and all design specifications are fully met.

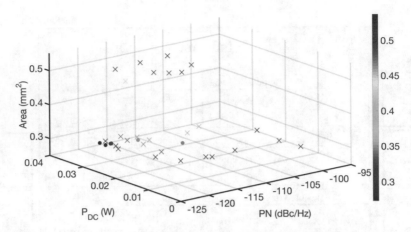

Fig. 5.10 run_{VCOS2} post-layout measures. Feasible points in post-layout are represented with circles (6 of 256) and unfeasible points (fail) are represented with crosses (27 of 256). The steady-state analysis failed convergence (n/c) in 223 points. (Reprinted with permission from [5])

5.3.1.3 Layout-Aware Optimizations

The previous results clearly justify the need for a layout-aware approach. Therefore, ten different runs of the layout-aware sizing optimization were performed following the strategy presented in Sect. 5.2. In contrast to the runs from Table 5.4, all device and interconnect parasitics are fully accounted for. The obtained results are presented in Table 5.6. Each optimization was conducted with a population of 64 individuals for 50 generations, from random initial solutions. Each layout-aware solution took on average 264 s to be evaluated on an Intel Core i7-3770 @ 3.4 GHz workstation with 32 GB of RAM, including complete layout generation, extraction, and simulation of the extracted netlist, which represents a significant overhead in terms of computational effort when compared to the 7 s required on average for pre-layout simulation only.

The POF obtained in the execution run_{VCOLA4} is shown in Fig. 5.12. A more detailed analysis of two sample points is presented in Table 5.7. The designs in these layout-aware POFs comply with the complete set of specifications, and therefore, are more trustworthy than the optimizations not considering full layout parasitics. Moreover, the inclusion of all required layout structures for manufacturing during optimization brings the designer closer to a first-pass fabrication success. The layout of a given point from the POF (referred to as LA1) is illustrated in Fig. 5.13.

5.3.1.4 Comparison Against Electromagnetic Simulation

Even though all stages of the proposed methodology were built over established CAD tools for RF IC design, in this Section an accuracy comparison is carried out between the proposed approach and the most accurate parasitic extractor available:

Table 5.5 Pre- and post-layout performance comparison of sample points from execution run_{VCOS2}

Performances	Specifications	Point S1[a]		Point S2		Point S3[a]	
		Schematic	Post-lay	Schematic	Post-lay	Schematic	Post-lay
f_{osc} (V_{tune} = 0 V)	2.55 GHz	2.556	**2.534**	2.596	2.572	2.666	2.629
f_{osc} (V_{tune} = 2.5 V)	2.45 GHz	2.360	2.330	2.408	2.378	2.427	2.379
PN@10 KHz	<−65 dBc/Hz	−75.61	−73.15	−76.30	−73.08	−72.26	−70.23
PN@100 KHz	<−92 dBc/Hz	−102.50	−99.29	−103.20	−99.33	−97.76	**−90.94**
PN@1 MHz	<−113 dBc/Hz	−124.54	−120.90	−124.40	−121.02	−119.05	**−111.01**
PN@10MHz	<−134 dBc/Hz	−144.00	−141.10	−144.60	−141.20	−138.40	**−131.00**
P_{DC}	<40 mW	27.44	23.28	26.57	20.06	15.07	14.62
V_{OUT}	0.15 V	1.874	1.339	1.831	1.202	0.978	0.465
Area	(Not constrained) mm²	0.2029	0.3005	0.2419	0.3483	0.1884	0.2803

[a]The performances in bold do not meet specifications

Fig. 5.11 Automatically generated layout (*a posteriori*) from the run_{VCOS2} (considering only sizing during the optimization). Layout with 0.348 mm² area. (Reprinted with permission from [5])

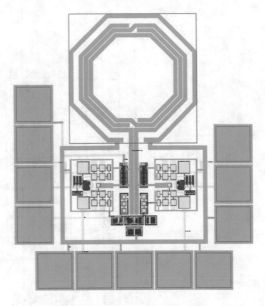

Table 5.6 Statistics of ten layout-aware optimizations with EM-characterized inductors

Run	Feasible solutions	Minimum and maximum objective values of the POF					
		Area (mm²)		PN@1 MHz (dBc/Hz)		P_{DC} (mW)	
		Min	Max	Min	Max	Min	Max
run_{VCOLA1}	14	0.3375	0.5123	−121.18	−116.30	13.019	32.968
run_{VCOLA2}	20	0.2764	0.3541	−121.40	−115.37	12.807	36.293
run_{VCOLA3}	14	0.3028	0.3867	−121.53	−114.89	14.300	31.002
run_{VCOLA4}	26	0.2734	0.3778	−121.76	−114.17	8.089	36.987
run_{VCOLA5}	8	0.3416	0.3716	−121.27	−118.81	26.350	35.193
run_{VCOLA6}	25	0.4535	0.6038	−122.57	−115.76	11.475	31.529
run_{VCOLA7}	20	0.3367	0.3642	−121.66	−117.40	21.729	29.357
run_{VCOLA8}	34	0.2841	0.3843	−121.00	−114.41	15.030	37.160
run_{VCOLA9}	38	0.3608	0.6207	−122.29	−114.01	14.269	39.840
$run_{VCOLA10}$	28	0.3453	0.5770	−122.41	−115.55	7.697	32.303

the EM simulator. To do so, the GDSII data from the layout point LA1 was imported in ADS *Momentum*. An EM simulation was performed to consider all the parasitic components of the routing and circuit shielding. The S-parameters obtained from the EM simulation were then back-annotated in the netlist in the same fashion as the inductors (i.e., *nport* in SpectreRF, or equivalently *Fblock* in EldoRF), and, the entire circuit simulated. The results of this simulation are shown in the last column of Table 5.7. Despite some minor differences between the measured post-layout performances resultant from the Calibre xRC extraction and the EM simulation, it is possible to observe that the post-layout simulation is very similar to the EM-extracted simulation, endorsing, therefore, the proposed approach.

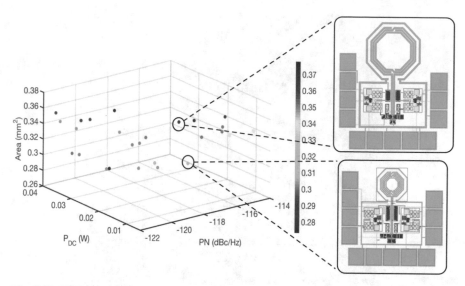

Fig. 5.12 VCO layout-aware optimization: Pareto front with the trade-off circuit area vs. PN@1 MHz vs. P_{DC} of RUN_{VCOLA4}. Two different layouts are also shown with at the same scale in order to illustrate the differences on the area. Point LA1 with higher area and point LA2 with lower area

Table 5.7 Performances of two designs from execution run_{VCOLA4}. Comparison between the performances for LA1 design when the routing is extracted with Calibre xRC or EM simulated

Performances	Specifications	LA2 design	LA1 design (Calibre xRC)	LA1 design (EM extracted)
f_{osc} ($V_{tune} = 0$ V)	2.55 GHz	2.558	2.632	2.648
f_{osc} ($V_{tune} = 2.5$ V)	2.45 GHz	2.272	2.347	2.373
PN@10 KHz	<-65 dBc/Hz	-72.39	-73.50	-73.09
PN@100 KHz	<-92 dBc/Hz	-97.58	-99.79	-100.28
PN@1 MHz	<-113 dBc/Hz	-118.71	-121.50	-122.97
PN@10 MHz	<-134 dBc/Hz	-138.80	-141.70	-142.45
P_{DC}	<40 mW	35.19	28.35	27.34
V_{OUT}	0.15 V	1.070	1.373	1.639
Area	(Not constrained) mm^2	0.2734	0.3512	0.3512

Although this routing EM simulation is the most accurate parasitic extraction method, it is a time-consuming process and a time-prohibitive solution to evaluate all candidate solutions within the optimization process. In this example, the EM simulation of a single routing layout took approximately 71 h to conclude for a setup of 700 frequency points and a mesh density of 30 cells/wavelength. An EM simulation with less frequency points and a wider mesh could be used to speed up the process, however, by using less frequency points the interpolation could no longer be passive and causal, and, therefore, no convergence is achieved during the

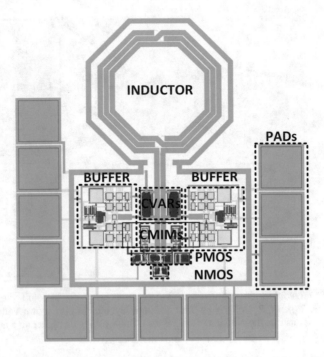

Fig. 5.13 Automatically generated layout annotated. Solution obtained from the run_{VCOLA4} (layout-aware optimization). Layout LA1, with 0.3512 mm^2 area

circuit simulation. On the other hand, by reducing the number of cells/wavelength, the price to pay is accuracy degradation.

5.3.2 LNA Design Using a Bottom-Up Systematic Layout-Aware Methodology

The previous Section clearly indicated the need for a layout-aware optimization. In this Section, the same methodology presented above, and applied to the design of a VCO, is going to be applied to the design of an LNA. The LNA topology is the same one used in Chap. 4, and its schematic representation is reported here for convenience (see Fig. 5.14).

The inductor topology chosen for this circuit is the octagonal asymmetric inductor with guard ring (shown in Fig. 5.15a), and the design variables of the inductor used to generate the POF are shown in Table 5.8. A multi-objective optimization was performed with 1000 individuals and 80 generations, resulting in the POF shown in Fig. 5.15b.

Fig. 5.14 Schematic of the
LNA

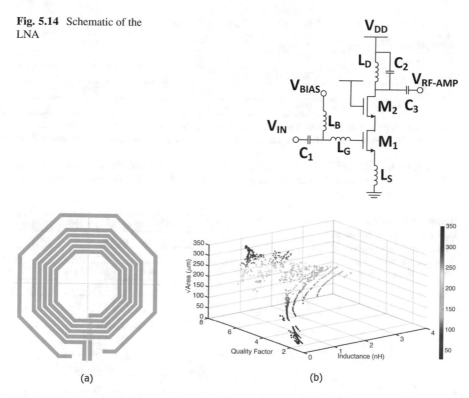

(a)   (b)

Fig. 5.15 Inductor topology and POF used in the LNA layout-aware optimization. (**a**) Instance
of the octagonal asymmetric inductor with guard ring. (**b**) POF of 1000 points with the trade-off
$\sqrt{\text{Area}}$ vs. Quality factor vs. Inductance of the octagonal asymmetric inductor with guard ring

Table 5.8 Inductor design
variables

Parameter	Minimum	Grid	Maximum
N	1	1	8
D_{in} (µm)	10	1	300
w (µm)	5	0.05	25
s (µm)	2.5	–	2.5

The circuit design variables and the optimization objectives and constraints are
shown in Tables 5.9 and 5.10, respectively. The circuit is intended to operate at
$V_{dd} = 2.5$ V.

For the LNA, the placement template hierarchy of Fig. 5.16 was defined taking
into account all the previous considerations inherent to RF design mentioned in
the previous Section and applied to the VCO layout-aware optimization. However,
since in the presented LNA, several inductors are used, a novel feature of the
methodology will be presented here. It was stated in Sect. 5.2.1, that one of the
RF improvements to AIDA-L was the ability to define *no-crossing* boxes, which are
used to keep consistent distances in all four directions during automatic placement.

Table 5.9 Design variables of the LNA optimization (other than inductors)

Design variables	Minimum	Maximum	Grid
w_{M1}, w_{M2} (μm)[a]	10	200	10
l_{M1}, l_{M2} (μm)	Fixed @ 0.35 μm		
V_b (V)	0	1.5	0.001
C_1, C_2, C_3[b] (μm)	9	29	1

[a]Since the foundry model transistor width is only valid up to 200 μm, three parallel transistors are used for both M_1 and M_2 in order to increase the ranges. The min and max value shown are for each transistor

[b]MIM capacitors are used for C_1, C_2, and C_3. Since the foundry model only allows values up to 29 μm, in order to increase the ranges, four parallel capacitors are used. The min and max value shown are for each capacitor

Table 5.10 Desired performances and constraints for the LNA layout-aware optimization with objectives Gain vs. Area vs. Power

LNA performances	LNA specifications
S_{11}	<-10 dB
S_{22}	<-10 dB
S_{21}	Maximize
S_{21}	>10 dB
k	>1
IIP_3	>-10 dBm
NF	<4 dB
Area	Minimize
P_{DC}	Minimize
P_{DC}	<25 mW

These *no-crossing* boxes are used in order to minimize coupling between devices, and, therefore, they will be used in the LNA template to minimize the coupling between the inductors, as it can be seen in Fig. 5.16. Although the inductors already have a guard ring, which allows for coupling minimization, by setting the inductors apart, this coupling is further minimized.

An optimization was performed with 128 individuals and 50 generations. The LNA POF obtained is shown in Fig. 5.17.

Two of the obtained designs are shown in detail in Table 5.11 and their layouts are shown in Figs. 5.18 and 5.19 (with the latter being annotated).

5.3.3   Mixer Design Using a Bottom-Up Systematic Layout-Aware Methodology

In this Section the layout-aware methodology previously described is applied to the design of a mixer. The topology considered for the experiments is shown in Fig. 5.20a. It is a Gilbert cell mixer, intended to down-convert the RF frequency from 2.46 GHz to 40 MHz. The testbench used for the mixer optimization is shown

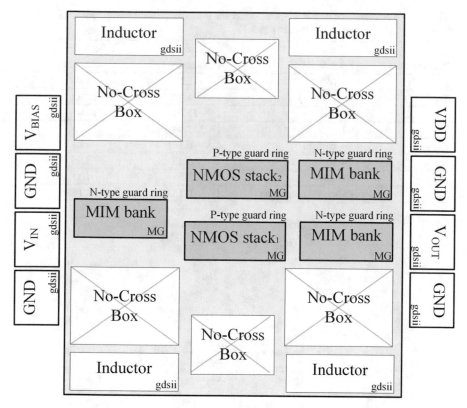

Fig. 5.16 Placement template hierarchy for the LNA. Shaded boxes refer to sub-partitions. Devices, blocks, or sub-partitions are either generated using the custom module generator or imported as a GSDII file (gdsii). Illustrating the use of *no-crossing* boxes in placement templates

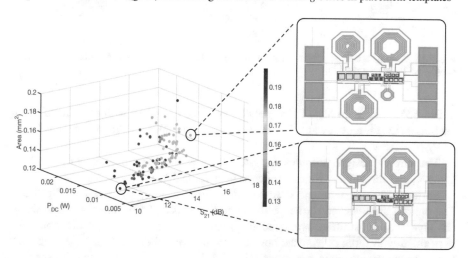

Fig. 5.17 POF obtained with the Layout-aware optimization of the LNA

Table 5.11 Performances of two LNA obtained for the layout-aware optimization with objectives Gain vs. Area vs. Power shown in Fig. 5.17

Performances	Specifications	LA1 design (highest gain)	LA2 design (lowest power)
S_{11}	$<-10\,dB$	-12.15	-12.40
S_{22}	$<-10\,dB$	-14.81	-11.90
S_{21}	$>10\,dB$	17.242	10.69
k	>1	537.04	378.03
IIP_3	$>-10\,dBm$	-0.141	-3.84
NF	$<4\,dB$	3.32	3.86
P_{DC}	$<25\,mW$	22.8	4.31
Area	(Not constrained) mm^2	0.1529	0.1588

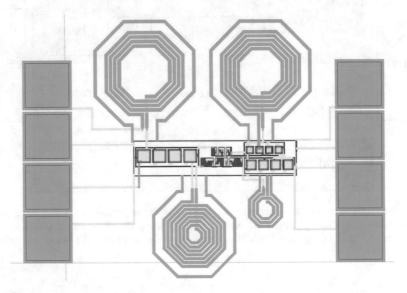

Fig. 5.18 Automatically generated layout from the LNA layout-aware optimization. Layout with the lowest P_{DC} from the POF

in Fig. 5.20b where an ideal RF signal at 2.46 GHz was set as well as an ideal local oscillator (LO) frequency at 2.5 GHz. The mixer will operate with a supply voltage of $V_{DD} = 2.5$ V.

The circuit design variables are the same as the experiments performed in Chap. 4, and shown here for convenience (see Table 5.12).

Furthermore, the template used for the optimization can be seen in Fig. 5.21. The optimization was performed with 128 individuals and 50 generations following the layout-aware strategy presented in previous Sections. The optimization was performed with three objectives: power consumption minimization, NF minimization, and conversion gain maximization. The desired specification can be seen in Table 5.13. In this optimization, in order to speed up the process, the CG is only

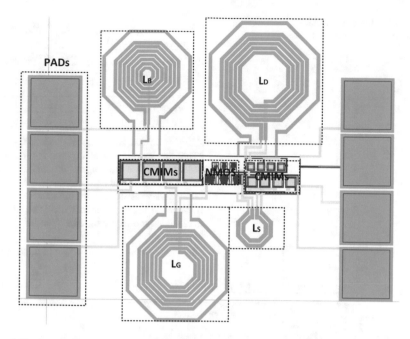

Fig. 5.19 Automatically generated layout from the LNA layout-aware optimization. Layout with the highest S_{21} from the POF

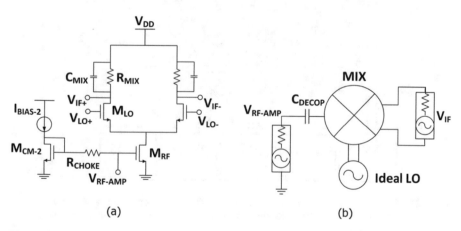

Fig. 5.20 (a) Schematic of the mixer and (b) its testbench

calculated at two IF frequency points: 10MHz and 40MHz. The same applies to the NF calculation. The optimization results can be observed in Fig. 5.22.

Two of the obtained designs are shown in detail in Table 5.14 and their layouts are shown in Figs. 5.23 and 5.24 (with the latter being annotated).

Table 5.12 Design variables for the mixer optimization

Variables	Min	Max	Grid
$w_{LO,RF,CM-2}$ (μm)	10	200	10
$l_{LO,RF,CM-2}$	Fixed @ 0.35 μm		
I_{BIAS-2} (mA)	0.1	1.5	0.1
$R_{W_{CHOKE,MIX}}$ (μm)[a,b]	1	3	1
$R_{l_{CHOKE,MIX}}$ (μm)[a,b]	3	90	1
C_{MIX} (pF)	0.3	3	0.3
C_{DECOP}	Fixed @ 10 pF		

[a]Three series resistances are used for R_{MIX} in order to increase the ranges. The min and max values shown are for each resistor

[b]Four series resistances are used for R_{CHOKE} in order to increase the ranges. The min and max values shown are for each resistor

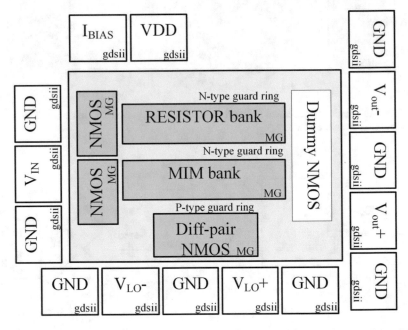

Fig. 5.21 Placement template hierarchy for the mixer. Shaded boxes refer to sub-partitions. Devices, blocks, or sub-partitions are either generated using the custom module generator or imported as a GSDII file (gdsii)

5.4 Optimization-Based Synthesis with Layout and Process Variability Considerations

In Sect. 4.3 a methodology was presented in order to include process variability during the optimization flow in order to achieve corner-aware POFs. Furthermore, in Sect. 5.3 a methodology was described which generates circuit layouts during the

Table 5.13 Design specifications for the mixer optimization

Mixer performance	Mixer specifications
CG @ 10 MHz	>0 dB
P_{DC}	Minimize
P_{DC}	<25 mW
NF @ 40 MHz	Minimize
NF @ 40 MHz	<20 dB
IIP_3	>0 dBm
Port-to-Port isolation	<−30 dB
Area[a]	–

[a]The area of the mixers is not considered for the optimization because this circuit does not have inductors. Therefore its area is relatively small when compared to the VCO or the LNA

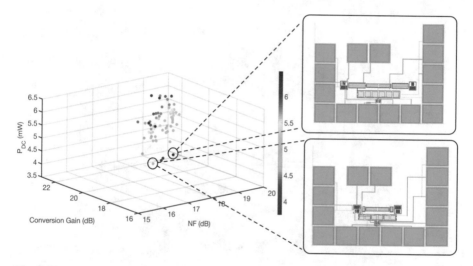

Fig. 5.22 POF obtained with the Layout-aware optimization of the mixer

optimization flow and considers all layout parasitics in order to achieve layout-aware POFs.

However, in order to achieve circuits which are truly suitable for a first-pass fabrication success, these methodologies have to be merged in order to deploy layout-corner-aware POFs. In such a methodology, both the process variations and the layout parasitics are taken into account during the optimization flow. The methodology is illustrated in Fig. 5.25. The main idea is to perform the optimization in two steps: first run a sizing optimization considering the typical and the device corner performances, using a corner-aware optimization, and afterwards, in a second step, the optimization is continued considering also the layout parasitics, performing a layout-corner-aware optimization. In this second step, the complete circuit layout is performed during the optimization, its parasitics are extracted and the circuit

Table 5.14 Performances of two mixer obtained for the layout-aware optimization with objectives NF vs. CG vs. Power shown in Fig. 5.22

Performances	Specifications	LA1 design (lowest NF)	LA2 design (lowest P_{DC})
CG @ 40 MHz	>0 dB	16.37	21.53
P_{DC}	<25 mW	5.64	3.74
NF @ 40 MHz	<20 dB	15.35	19.42
IIP_3	>0 dBm	10.68	7.62
RF-IF isolation	<−30 dB	−35.08	−35.76
LO-IF isolation	<−30 dB	−36.34	−39.54
Area	(Not constrained) mm^2	0.0779	0.0805

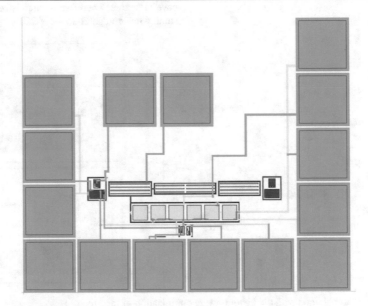

Fig. 5.23 Automatically generated layout from the mixer layout-aware optimization. Layout with the lowest P_{DC} from the POF

performances are evaluated with a commercial simulator. However, during this second step, the process variability is also taken into account by simulating the extracted layout considering its device corner performances. In this Section, a VCO is optimized using such layout-corner-aware methodologies in order to illustrate such methodology.

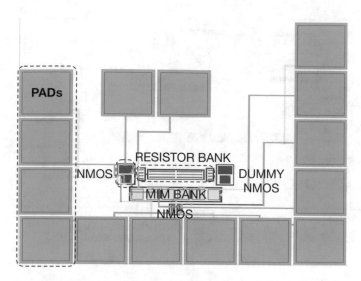

Fig. 5.24 Automatically generated layout from the mixer layout-aware optimization. Layout with the lowest NF from the POF

5.4.1 VCO Design Using a Bottom-Up Systematic Layout-Corner-Aware Methodology

In this Section, a VCO is designed using the previously described layout-corner-aware methodology. The circuit topology is the one previously presented in Fig. 5.6. The inductor topology chosen for the circuit optimization and its respective POF are shown in Fig. 5.7a, b, respectively. The circuit design variables and the optimization objectives and constraints are shown in Tables 5.2 and 5.3, respectively. The circuit is intended to operate at $V_{dd} = 2.5$ V. A corner-aware optimization was performed with 128 individuals and 100 generations, and afterwards, the layout-corner-aware optimization was started, with 128 individuals and 30 generations. The obtained POF is shown in Fig. 5.26. The performances of the individuals with lowest area and phase noise are detailed in Table 5.15. Furthermore, the layouts of these two solutions are shown in Figs. 5.27 and 5.28. When compared with the VCOs obtained only with the layout-aware VCO POF, it can be seen that the area is usually higher in the POF which complies with the constraints in the extreme corner performances. This fact is compliant with the conclusion obtained in Sect. 4.3, when only the sizing of the circuits were considered, but the process variability was also taken into account.

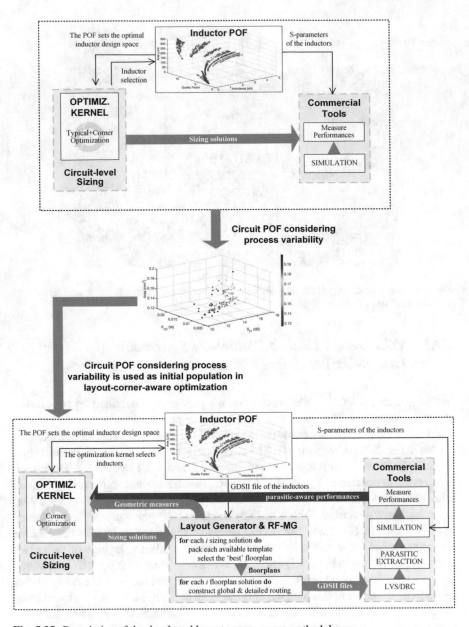

Fig. 5.25 Description of the developed layout-corner-aware methodology

5.5   Summary

In this chapter a layout and variability-aware optimization-based approach has been described which has the capability to include parasitics and variability in

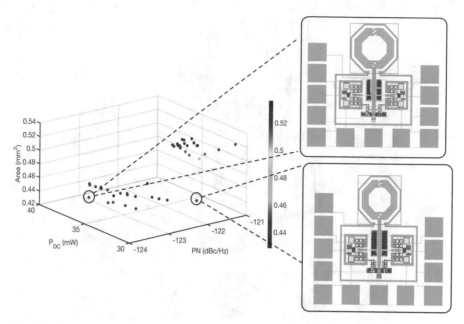

Fig. 5.26 Layout-Corner-aware optimization of the VCO

Table 5.15 Performances of two points from the optimization illustrated in Fig. 5.26

Specifications		Point LA1 (lowest area)	Point LA2 (lowest phase noise)
f_{osc} ($V_{tune} = 0$ V)	>2.55 GHz	2.5938	2.5503
f_{osc} ($V_{tune} = 2.5$ V)	<2.45 GHz	2.4042	2.4438
PN@10 KHz	<−65 dBc/Hz	−66.96	−73.30
PN@100 KHz	<−92 dBc/Hz	−96.10	−100.94
PN@1 MHz	Min <−113 dBc/Hz	−121.6	−123.6
P_{DC}	<40 mW	38.34	36.22
V_{OUT}	0.15 V	1.70	1.50
Area	(Not constrained) mm^2	0.0428	0.0446

the optimization loop. Moreover, the approach has been developed with the intent to exploit the full capabilities of well-known and established off-the-shelf CAD tools. Promising results over widely used RF circuits (LNA, VCO, and mixer) have been provided as a proof-of-concept and validated. By using such approach, design iterations can be successfully eliminated, bringing these RF circuit blocks closer to a first-pass fabrication success as every structure required for tape-out is considered and balanced during the optimization process.

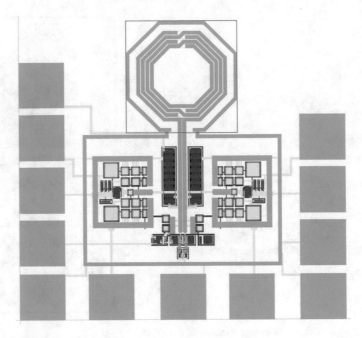

Fig. 5.27 Point LA1. VCO with the lowest area available from the layout-corner-aware optimization shown in Fig. 5.26

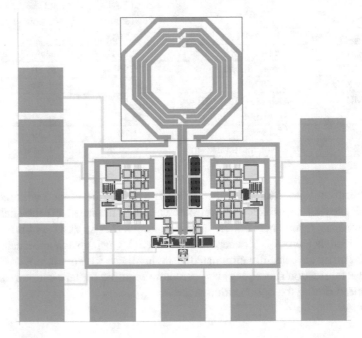

Fig. 5.28 Point LA2. VCO with the lowest phase noise available from the layout-corner-aware optimization shown in Fig. 5.26

References

1. R. Martins, N. Lourenço, S. Rodrigues, J. Guilherme, N. Horta, AIDA: automated analog IC design flow from circuit level to layout, in *International Conference on Synthesis, Modeling, Analysis and Simulation Methods and Applications to Circuit Design* (2012), pp. 29–32
2. R. Martins, Placement, Routing and Parasitic Extraction Techniques applied to Analog IC Design Automation. Ph.D. Thesis, Universidade de Lisboa, Instituto Superior Tecnico (2015)
3. R. Martins, N. Lourenço, N. Horta, LAYGEN II: automatic layout generation of analog integrated circuits. IEEE Trans. Comput. Aided Des. Integr. Circuits Syst. **32**, 1641–1654 (2013)
4. R. Martins, N. Lourenço, A. Canelas, N. Horta, Electromigration-aware and IR-drop avoidance routing in analog multiport terminal structures, in *Design, Automation & Test in Europe Conference & Exhibition* (2014), pp. 1–6
5. R. Martins, N. Lourenço, F. Passos, R. Póvoa, A. Canelas, E. Roca, R. Castro-López, J. Sieiro, F. V. Fernández, N. Horta, Two-step RF IC block synthesis with preoptimized inductors and full layout generation in-the-loop. IEEE Trans. Comput. Aided Des. Integr. Circuits Syst. **38**, 989–100 (2019)
6. Calibre, https://www.mentor.com/training/course_categories/calibre. Accessed 28 Jan 2020

Chapter 6
Multilevel Bottom-Up Systematic Design Methodologies

In this chapter the bottom-up design methodology described in Chap. 4 will be applied to the design of an RF front-end receiver. With such methodology, the passive level is first designed (e.g., several inductors topologies), then the circuit level (e.g., LNA, VCO, and mixer), until reaching the system level, where all the circuits are connected in order to design the RF front-end receiver. This was the first time that such bottom-up multilevel methodology was used for RF circuits, from the device up to the front-end level [1].

Since several blocks will be connected, in Sect. 6.1 some brief considerations are given on cascaded RF systems. While designing an RF front-end for a given communication standard, the designer must impose the needed receiver specifications (e.g., NF or IIP_3). However, these are not normally specified by the communication standards. Therefore, in order to design the receiver, one must derive the system specifications from the communication standards. In Sect. 6.2, a brief explanation on how can this be performed is given. In Sect. 6.3, the bottom-up design of the RF receiver is performed.

It was demonstrated in Sect. 4.2 that, between the circuit and the passive level, the hierarchical partitioning and bottom-up hierarchical composition show several advantages when compared to techniques where such hierarchical partitioning is not performed. Therefore, it is also interesting to make this comparison at the system level. Therefore, in Sect. 6.4, different hierarchical partitioning strategies are applied to the design of the RF front-end and their efficiency is compared.

6.1 Cascaded Radio-Frequency Systems

When designing a complete RF system, the designer needs to have a clear insight on how the performances of each block influence the entire system. Therefore, a brief study on such subject is performed in this section. One of the primary performance

© Springer Nature Switzerland AG 2020
F. Passos et al., *Automated Hierarchical Synthesis of Radio-Frequency Integrated Circuits and Systems*, https://doi.org/10.1007/978-3-030-47247-4_6

measures of an RF front-end used in, e.g., digital communications, is the bit error rate (BER), which is related to the SNR. The specifications on these two parameters (BER and SNR) will determine the lower-level specifications of individual blocks, such as gain (G), NF, and IIP_3. While designing the RF system, these three performances must be intelligently chosen because they are affected by other blocks in the RF chain.

6.1.1 Gain of Cascaded Stages

Consider, for instance, the cascaded system in Fig. 6.1, consisting of three different RF stages.

The total gain of the cascaded system in Fig. 6.1 is given by

$$G_{total} = 20\log_{10}(G_1) + 20\log_{10}(G_2) + 20\log_{10}(G_3) \tag{6.1}$$

For a cascaded system with i stages, the total gain would be

$$G_{total} = 20\log_{10}(G_1) + 20\log_{10}(G_2) + 20\log_{10}(G_3) + \ldots + 20\log_{10}(G_i) \tag{6.2}$$

6.1.2 Noise Figure of Cascaded Stages

Again, consider the cascaded system shown in Fig. 6.1. The NF of a cascaded system was approximated by Friis in [2]. The NF of the cascaded system in Fig. 6.1 is given by

$$NF_{total} = NF_1 + \frac{NF_2 - 1}{G_1} + \frac{NF_3 - 1}{G_1 G_2} \tag{6.3}$$

For a cascaded system with i stages, the total NF would be

$$NF_{total} = NF_1 + \frac{NF_2 - 1}{G_1} + \frac{NF_3 - 1}{G_1 G_2} + \cdots + \frac{NF_i - 1}{G_1 G_2 \ldots G_{i-1}} \tag{6.4}$$

Fig. 6.1 RF cascaded system

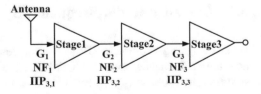

It has to be taken into account that the original Friis formulas were derived for values of NF in linear scale and not their values in dB. Therefore, in order to get a NF value in dB, the following formula must be used:

$$NF_{total}(dB) = 10 \log_{10} \left(NF_1 + \frac{NF_2 - 1}{G_1} + \frac{NF_3 - 1}{G_1 G_2} \right) \tag{6.5}$$

From Eq. (6.4), it is possible to conclude that the first stage on the RF chain has the highest influence on the entire system, because it is the highest contributor to the total noise.

6.1.3 IIP_3 of Cascaded Stages

In a similar fashion as for the NF, we can evaluate the third-order intercept point of a cascaded system. However, the production of intermodulation distortion in a cascaded system is somewhat more complicated than the NF, due to the distortion products in each stage, which may have arbitrary phase relationships that make it difficult to precisely determine the cumulative distortion. However, it is possible to reach a simplifying (and somewhat optimistic) formula, given by

$$\frac{1}{IIP_{3,total}} \approx \frac{1}{IIP_{3,1}} + \frac{G_1}{IIP_{3,2}} + \frac{G_1 G_2}{IIP_{3,3}} \tag{6.6}$$

For a cascaded system with i stages, the total IIP_3 would be

$$\frac{1}{IIP_{3,total}} \approx \frac{1}{IIP_{3,1}} + \frac{G_1}{IIP_{3,2}} + \frac{G_1 G_2}{IIP_{3,3}} + \cdots + \frac{G_1 G_2 \ldots G_i}{IIP_{3,i}} \tag{6.7}$$

From Eq. (6.7), it is possible to conclude that the last stage of the RF chain has the highest influence on the entire system, because it is the highest contributor of nonlinearities.

It is possible to see from Eqs. (6.4) and (6.7) that a natural tapering occurs in cascade systems, where early stages contribute more to the NF and later stages contribute more to the IIP_3, as illustrated in Fig. 6.2.

6.1.4 Inter-Block Connection

In this work, each circuit is optimized independently, but will afterwards be connected to another (e.g., LNA output is connected to the mixer input, as seen in Fig. 6.3a), which might slightly change its performances. In IC technologies (up to a given frequency), where the routing interconnection between circuits is

Fig. 6.2 The relative
contribution of successive
stages to noise and distortion

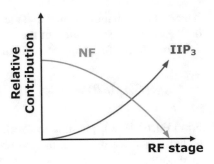

rather small, the concept of reflected wave does not apply entirely [3]; however, the input/output impedance of each circuit should be taken into account, especially in RF signal paths such as the interconnection between the LNA and the mixer. The typical input capacitance of a NMOS transistor in the adopted technology (0.35 μm CMOS technology) can go from 50 fF up to 200 fF (considering the minimum and maximum unitary transistor sizes). Therefore, in order to show how the performances of, e.g., the LNA are affected due to the change in this capacitance, a parametric analysis was performed, where a capacitance was connected to the output of the LNA (see C_{OUT} in Fig. 6.3b). This capacitance was then varied from 50 fF up to 200 fF. The results can be shown in Fig. 6.4. It is possible to observe that the NF and LNA gain (S_{21}) performances are barely affected by the capacitance C_{OUT}. It is clear in Fig. 6.4b that the output matching (S_{22}) is the performance that mostly changes due to this capacitance. However, the LNA still presents good output matching (< -12 dB) in the desired band. This means that whatever the size of the input mixer transistor (M_{RF} in Fig. 6.3d), the LNA will still present the desired performances. Nevertheless, for a more accurate design, these impedances at the output and input of connected blocks should still be accounted for, as in the testbench of Fig. 6.3b, c, where C_{OUT} is set to be 100 fF.

6.2   From Communication Standards to Circuit Specifications

Usually, the key specifications for a receiver are the NF and the IIP_3. However, these specifications are not directly specified by the standards, but rather have to be derived from them. Therefore, a brief study on how to do so is presented in this section. One of the primary standard specifications is the sensitivity of the receiver, which is the minimum input power signal level that is required to achieve a certain output bit error rate (BER), which is usually also defined by the standard. Furthermore, to achieve the specified BER, a minimum signal-to-noise ratio (SNR_{min}) must be achieved. The SNR-BER curve depends on the modulation used for the data transmission and can be calculated by any high-level specification tool, e.g., bit error rate toolbox in MATLAB. The required receiver NF is given by

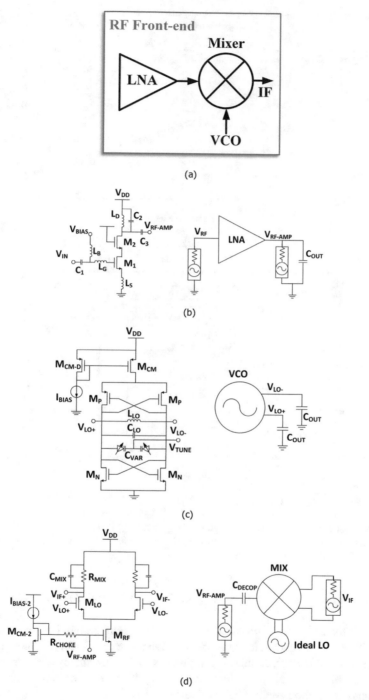

Fig. 6.3 (**a**) Front-end (**b**) LNA topology and testbench (**c**) VCO topology and testbench (**d**) Mixer topology and testbench

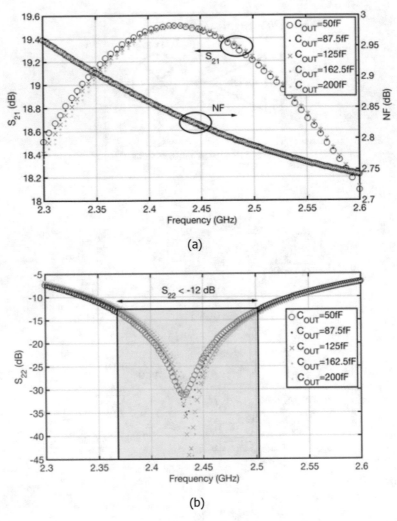

Fig. 6.4 Illustrating how (**a**) the noise figure (NF), gain (S_{21}) and (**b**) the output matching (S_{22}) of the LNA are affected by the change in C_{OUT}. The shaded area in (**b**) represents the entire frequency band where $S_{22} < -12\,\text{dB}$

$$NF = sensitivity - SNR_{min} - 10\log(kT) + 10\log(BW) - margin \qquad (6.8)$$

where k is the Boltzmann constant, T is the absolute temperature (in Kelvin degrees) and BW is the channel selection filter bandwidth. Several dBs of margin can be added for possible degradations due to process variations, mismatches or any other effects that are not taken into account in Eq. (6.8). The IIP_3, which accounts for the nonlinearities in the system, can be calculated by

$$IIP_3 = \frac{1}{2}(3P_{int} - P_{sig} + SNR_{min}) - margin \qquad (6.9)$$

where P_{int} and P_{sig} are the interferer and signal levels, respectively, when the intermodulation test is performed. These signal levels and their frequencies are also defined by the standard. Again, several dBs of margin are also added.

6.3 RF Front-End Systematic Design

It was shown in Chap. 4 that bottom-up design strategies are more efficient and achieve better results than optimization-based methodologies with no hierarchical decomposition. There are several other motivations to use bottom-up design methodologies. In bottom-up design methodologies, while going up the hierarchy, the lower levels are reduced to an optimal search space. This means that when designing a given high-level circuit or system, the optimization algorithm is only performing an exploration in an optimal design space, which helps improving the efficiency of the entire process, because the algorithm no longer has to search in unusable/suboptimal design areas.

In Fig. 6.5, the entire proposed strategy can be observed. In such figure, it is possible to observe that the design stage starts from the lowest possible level: device level. Here, any given inductor topology can be optimized using a multi-objective optimization algorithm in order to reach a POF. In this book, the SIDe-O tool is used. Afterwards, in order to establish the search space for higher-level optimizations, the inductor POFs have to be mapped into matrices, as explained in Chap. 4. After mapping the inductor POFs, it is possible to start the circuit-level optimizations. In this case, in order to form the RF front-end, an LNA, VCO, and mixer are needed. After the three circuit optimizations are finished, all the circuit POFs are mapped into matrices and can then be used in system-level optimizations in order to build the entire circuit.

Another motivation to use bottom-up design methodologies is that it encourages the hierarchical reusability of lower-level blocks. Since in bottom-up design methodologies the lower-level blocks are first designed, the obtained POFs can be stored and used afterwards in the composition of any other system. This means that for the front-end design depicted in Fig. 6.3a, the inductor topologies and each individual circuit (LNA, VCO, and mixer) only have to be optimized once. Afterwards, the front-end can be optimized for other communication standards, without having to perform any low-level optimization (as long as the operation frequency is maintained). Again, this highly improves the efficiency of the methodology.

Moreover, by using such methodology, it is also possible to consider several different circuit topologies for each low-level circuit. Thus, all levels of the bottom-up chain can be implemented with different topologies and, therefore, while performing an actual system synthesis, the evolutionary optimizer can combine not only different circuit designs but also different circuit topologies (all selected from

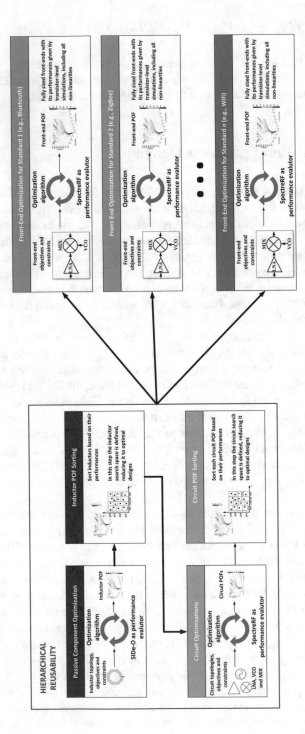

Fig. 6.5 Illustrating the proposed design methodology for the design of RF systems, applied to the specific case of an RF front-end and also the hierarchical reusability of the methodology

pre-generated POFs) [4]. The above-mentioned characteristics provide the basis for an efficient and highly dynamic system synthesis, as the low-level optimizations only have to be performed once and a number of circuit topologies can be used. This hierarchical reusability and the entire proposed design methodology applied to the specific case of an RF front-end are illustrated in Fig. 6.5. It should be said that, for the case of RF systems, the limitation for this hierarchical reusability is the frequency band, since each passive/circuit is optimized for a given frequency band.

6.3.1   Passive Components and Individual Circuits Synthesis

6.3.1.1   Symmetric and Asymmetric Octagonal Inductors

The first step of the methodology is to obtain the passive component POFs. In this chapter, two different inductor topologies are considered: an octagonal asymmetric topology (used in the LNAs) and an octagonal symmetric topology (used in the VCOs). These optimizations are the same performed in Chap. 4 for the optimization of the LNA and the VCOs. However, the inductors topologies and POFs are shown in Fig. 6.6 for convenience. For the same reason, the search space for the inductors is presented in Table 6.1.

The POFs were obtained with SIDe-O and had three objectives: maximization of the quality factor, Q, and inductance, L, while minimizing the area. The optimizations were performed with 1000 individuals and 80 generations and the inductors were subject to the same constraints as presented in Chap. 3. The obtained POFs can be seen in Fig. 6.6. After the inductor optimization is performed, the inductor POF is indexed into a matrix and used as inductor search space in the higher-level circuit optimizations.

6.3.1.2   Low Noise Amplifier

The LNA is intended to operate at any frequency of the ISM band (2.4–2.5 GHz), with a supply voltage V_{DD}=2.5 V. The LNA topology considered is a source degenerated LNA shown in Fig. 6.3b.

It was discussed in Sect. 6.1 that the NF of the first block in the RF chain is very important for the overall NF of the system. Therefore, the main concern while designing the LNA is to obtain a very low NF. On the other hand, the IIP_3 of the LNA is the one that least influences the overall performance of the front-end. Nevertheless, its value is still important for the front-end design. The LNA optimization was performed with four objectives: S_{21} maximization and area, NF, and P_{DC} minimization. The optimization settings can be seen in the second column, top row, of Table 6.2. The desired specifications (optimization constraints) can be seen in the first and second column of Table 6.2, the design variables can be seen in Table 6.3 and the results of the optimization can be seen in Fig. 6.7. In this

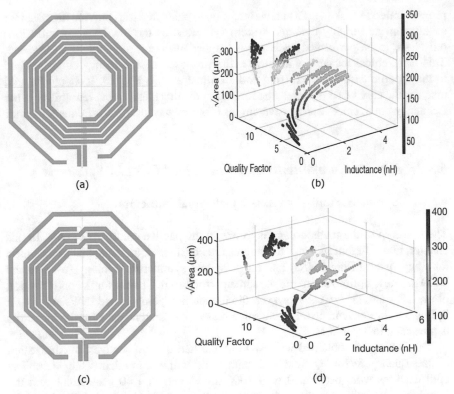

Fig. 6.6 Inductor topologies and POFs used in the VCO and LNA optimizations. (**a**) Octagonal asymmetric inductor topology. (**b**) POF of 1000 points with the trade-off √Area vs. Quality factor vs. Inductance of the octagonal asymmetric inductor. (**c**) Octagonal symmetric inductor topology. (**d**) POF of 1000 points with the trade-off √Area vs. Quality factor vs. Inductance of the octagonal symmetric inductor.

Table 6.1 Inductor design variables

Parameter	Minimum	Grid	Maximum
N	1	1	8
D_{in} (μm)	10	1	300
w (μm)	5	0.05	25
s (μm)	2.5	–	2.5

POF, the designer would have the best designs available for the selected trade-off, which means for a given value of S_{21}, the LNA with lowest NF, P_{DC}, and area is available in the POF. Consider that, unlike previous optimization examples, this one has four objectives. Therefore, for the graphical representation the three coordinate axes have been used for three objectives and a color code has been used for the fourth objective.

Table 6.2 Desired specifications for the LNA, VCO, mixer optimizations

LNA optimization settings		VCO optimization settings		Mixer optimization settings	
Individuals	800	Individuals	300	Individuals	300
Generations	300	Generations	100	Generations	100
LNA performances	LNA specifications	VCO performances	VCO specifications	Mixer performances	Mixer specifications
S_{11} @ 2.45; 2.5; 2.55 GHz	<−12 dB	f_{osc}	>2.45 GHz	CG @ 10 MHz	>5 dB
S_{22} @ 2.45; 2.5; 2.55 GHz	<−12 dB	f_{osc}	<2.55 GHz	CG @ 40 MHz	>5 dB
S_{21} @ 2.45; 2.5; 2.55 GHz	Maximize	PN @ 1 MHz offset	Minimize	P_{DC}	Minimize
k	>1	P_{DC}	Minimize	NF @ 10 MHz	<20 dB
NF @ 2.45; 2.5; 2.55 GHz	Minimize	P_{OUT}	Maximize	NF @ 40 MHz	Minimize
P_{DC}	Minimize			IIP_3	Maximize
IIP_3	>−15 dBm			Port-to-port isolation	<−30 dB
S_{21} @ 2.45; 2.5; 2.55 GHz	>7 dB	PN @ 1 MHz offset	<−110 dBc/Hz		
		P_{OUT}	>5 dBm	NF @ 40 MHz	<20 dB
P_{DC}	<20 mW	P_{DC}	<20 mW	P_{DC}	<20 mW
				IIP_3	>−20 dBm
Inductors	From POF	Inductors	From POF	Inductors	–
Area	Minimize	Area	Minimize	Area	Minimize

Table 6.3 Design variables of the LNA optimization (other than inductors)

Design variables	Minimum	Maximum	Grid
w_{M1}, w_{M2} (μm)[a]	10	200	10
l_{M1}, l_{M2} (μm)	Fixed @ 0.35 μm		
V_b (V)	0	1.5	0.001
C_1, C_2, C_3^b (μm)	9	29	1

[a] Since the foundry model transistor width is only valid up to 200 μm, three parallel transistors are used for both M_1 and M_2 in order to increase the ranges. The min and max value shown are for each transistor
[b] Four parallel MIM capacitors are used for C_1, C_2, and C_3 in order to increase the ranges. The min and max value shown are for each capacitor

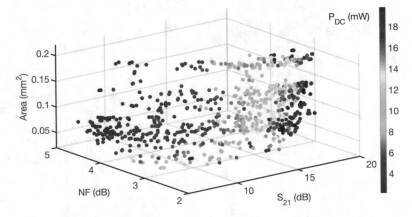

Fig. 6.7 POF obtained for the LNA. The color bar represents the power consumption, which is the fourth objective of the optimization. (Reprinted with permission [1])

6.3.1.3 Voltage Controlled Oscillator

The VCO is intended to oscillate at a frequency of 2.5 GHz with a supply voltage of V_{DD}=2.5 V. From the several VCO topologies available, in this work, a cross-coupled double-differential VCO is considered, as depicted in Fig. 6.3c. The optimization was performed with four objectives: maximization of P_{OUT} (which is the output swing measured in dBm) and minimization of PN, P_{DC}, and area. The optimization settings can be seen in the fourth column, top row, of Table 6.2. It is possible to observe that there is a difference between the number of individuals and generations of the LNA and the VCO. This difference is due to the fact that the LNA plays a major role in the front-end (e.g., for the NF and IIP_3 performances), has more complex trade-offs and a higher number of design constraints; therefore, its optimization is more complex and a POF with higher density is desired for the receiver optimization. Therefore, for further composition of the system, it is desired that the LNA has a higher number of design options, covering a wider design area. The desired specifications can be seen in the third and fourth column of Table 6.2,

Table 6.4 Design variables of the VCO (other than inductors)

Variables	Minimum	Maximum	Grid
w_{n1} (μm)	10	200	10
$w_{p1,d,dd}$ (μm)	10	150	10
$l_{n1,p1,d,dd}$ (μm)	Fixed @ 0.35 μm		
I_{bp} (mA)	0.1	1.5	0.1
w_{Cvar} (μm)	Fixed @ 6.6 μm		
l_{Cvar} (μm)	Fixed @ 0.65 μm		
Inductors	Selected from the POF		
Row_{Cvar}, Col^a_{Cvar}	4	12	1
C (μm)[b]	9	29	1

[a] Row_{Cvar} and Col_{Cvar} are the number of fingers per row and column of the varactors
[b] 6 MIM capacitors were used. The min and max values are for each one of the 6 MIM capacitors used

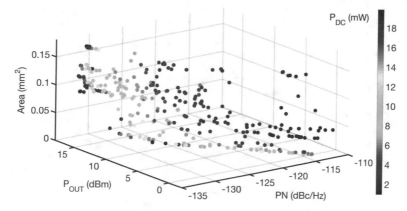

Fig. 6.8 POF obtained for the VCO. The color bar represents the power consumption, which is the fourth objective of the optimization. (Reprinted with permission [1])

the design variables can be seen in Table 6.4, and the optimization results can be seen in Fig. 6.8. Similarly to the LNA optimization, since 4 objectives are pursued, for the graphical representation the three coordinate axes have been used for three objectives and a color code has been used for the fourth objective.

6.3.1.4   Mixer

The mixer is intended to down-convert the RF frequency to the IF frequency. The Gilbert cell mixer topology shown in Fig. 6.3d is used.

The optimization was performed with four objectives: IIP_3 maximization and P_{DC}, NF, and area minimization. The optimization settings can be seen in the sixth column of Table 6.2. Again, the number of optimization individuals and generations

Table 6.5 Design variables for the mixer optimization

Variables	Min	Max	Grid
$w_{LO,RF,CM-2}$ (μm)	10	200	10
$l_{LO,RF,CM-2}$	Fixed @ 0.35 μm		
I_{BIAS-2} (mA)	0.1	1.5	0.1
$R_{W_{CHOKE,MIX}}$ (μm)[a,b]	1	3	1
$R_{l_{CHOKE,MIX}}$ (μm)[a,b]	3	90	1
C_{MIX} (pF)	0.3	3	0.3
C_{DECOP}	Fixed @ 10pF		

[a]Three series resistances are used for R_{MIX} in order to increase the ranges. The min and max values shown are for each resistor
[b]Four series resistances are used for R_{CHOKE} in order to increase the ranges. The min and max values shown are for each resistor

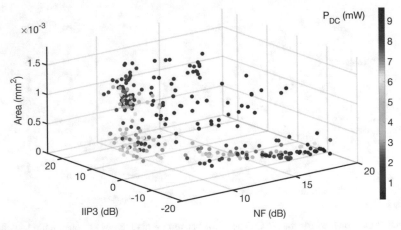

Fig. 6.9 POF obtained for the mixer. The color bar represents the power consumption, which is the fourth objective of the optimization. (Reprinted with permission [1])

were selected accordingly to the optimization problem. The desired specifications (optimization constraints) can be seen in the fifth and sixth column of Table 6.2, the design variables can be seen in Table 6.5, and the results of the optimization can be seen in Fig. 6.9. In Table 6.2, note that the mixer CG and NF constraints were imposed at two different IF frequencies in order to guarantee that the constraints are met for an IF band from 10 to 40 MHz.

6.3.2   Multilevel Bottom-up Receiver Front-End Synthesis

In this sub-section, the synthesis of the receiver front-end is performed in a bottom-up fashion, using the previously generated POFs from the LNA, VCO, and mixer.

Since the obtained circuit POFs have 4 objectives, these are indexed into 3D matrices which are then used as optimal design space for the receiver front-end synthesis. The front-end optimization is performed in order to join the individual blocks that together empower the best front-ends for a given communication standard. Due to the hierarchical POF reusability that the methodology enables, the previous optimizations (passives and circuits) only have to be performed once for a given frequency band. Afterwards, the POFs can be stored and reused for other communication standards that operate in the same frequency band. In the next sub-sections, a receiver synthesis is performed for two different standards, where the circuit specifications are derived using Eqs. (6.8) and (6.9) and the system is designed following the already discussed bottom-up design methodology.

6.3.2.1 Bluetooth

According to the Bluetooth standard (IEEE 802.15.1), which has a data rate of 720 Kbits/s, the required receiver sensitivity is -70 dBm for a BER= 10^{-3}. In order to evaluate the effect of non-idealities, we need to know the SNR_{min} required to meet the BER specifications. Baseband simulations in [5] showed that the required SNR_{min} is 12.25 dB using a Gaussian frequency shift keying (GFSK) modulation with a bandwidth per bit rate (BT) of 0.5 and a modulation index h=0.32. Although the minimum sensitivity is -70 dBm, many works have presented Bluetooth receivers with sensitivities better than -80 dBm [6, 7]. Hence, in this experiment we are improving the sensitivity down to -85 dBm.

From Eq. (6.8), and using a margin of 10 dB for a BW channel of 600 kHz, the maximum NF is 8.79 dB. For the intermodulation test, the Bluetooth standard defines that $P_{int} = -39$ dBm and $P_{sig} = -64$ dBm. Therefore, from Eq. (6.9), and using a margin of 10 dB, the minimum IIP_3 is -10.35 dBm. Furthermore, assuming that the receiver is for, e.g., an IoT application, the receiver area and P_{DC} need to be kept at a minimum. The constraints and objectives of the receiver synthesis are shown in Table 6.6, as well as the number of individuals and generations for the optimization. Two different optimizations (denoted as example 1 and example 2) were performed. The main difference between them is the conversion gain given as constraint: in example 1 a more relaxed constraint of 12 dB is imposed and in example 2 a high gain of 30 dB is required. The gain is not usually a specification of the standard, however, achieving higher conversion gains in the front-end can help relaxing the specifications for following blocks (e.g., analog-to-digital converter).

Similarly to the mixer optimization, note that the receiver CG and NF constraints were imposed at two different IF frequencies in order to guarantee that the constraints are met for an IF band. However, while in the mixer testbench an ideal LO of 2.50 GHz was used, for the receiver, a real VCO is used, which may oscillate between 2.45 and 2.55 GHz (as imposed in the VCO constraints). Therefore, the up- and down-frequency will vary depending on which VCO is used in the receiver. The result of the optimizations can be seen in Fig. 6.10. Each of the red and black circles in Fig. 6.10 represents a fully sized front-end, compliant with the Bluetooth standard

Table 6.6 Desired specifications for the receiver front-end optimizations and the number of individuals and generations for the optimization

	Example 1		Example 2	
Optimization settings	Bluetooth	Wi-Fi	Bluetooth	Wi-Fi
Individuals	160	160	160	160
Generations	60	60	60	60
Front-end performance	Bluetooth	WLAN	Bluetooth	WLAN
CG @ down-frequency	>12 dB	>12 dB	>30 dB	>30 dB
CG @ up-frequency	>12 dB	>12 dB	>30 dB	>30 dB
P_{DC}	<40 mW	<40 mW	<40 mW	<40 mW
P_{DC}	Minimize	Minimize	Minimize	Minimize
NF @ down-frequency	<8.79 dB	<5.64 dB	<8.79 dB	<5.64 dB
NF @ up-frequency	<8.79 dB	<5.64 dB	<8.79 dB	<5.64 dB
IIP_3	>−10.35 dBm	>−20.3 dBm	>−10.35 dBm	>−20.3 dBm
Area	Minimize	Minimize	Minimize	Minimize

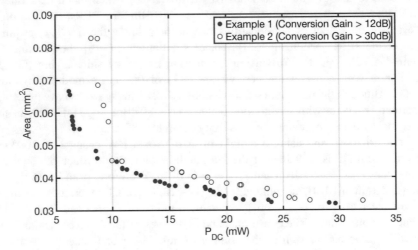

Fig. 6.10 POF of the front-ends compliant with the Bluetooth standard, for both examples in Table 6.6

with its performances estimated with device-level accuracy. The POF for example 1 is denoted by the red solid circles, and the POF for example 2 is denoted by black circles. It is possible to observe that for example 1, the front-ends designs have better power-area trade-offs, which is the effect of not imposing such a hard constraint in conversion gain. The designer may then select a design with appropriate area-power trade-off among all the possible choices offered at the final results.

6.3.2.2 Wi-Fi/WLAN Receiver Synthesis

The Wi-Fi/WLAN standard (IEEE 802.11b) has a data rate of 11 Mbit/s and the required receiver sensitivity is -76 dBm for a BER=10^{-5}. The 802.11b radio link uses a direct sequence spread spectrum technique called complementary code keying (CCK). The modulation technique used is the quadrature phase shift keying (QPSK) and the needed SNR_{min} to meet the required BER is 11.4 dB. In this experiment, we are pushing the sensitivity down to -79 dBm. From Eq. (6.8), using a margin of 10 dB and a channel BW of 6 MHz, the maximum NF is 5.64 dB. There is no standalone IIP_3 test for Wi-Fi/WLAN, and since the interferer is a wideband signal, the direct application of Eq. (6.9) could be inaccurate [5]. However, the required IIP_3 can be estimated. In [5] the required IIP_3 was derived specifically for the Wi-Fi standard using

$$IIP_3 = \frac{1}{2}\left(3P_{int} - 37 - P_{sig} + SNR_{min}\right) - margin \tag{6.10}$$

The interference and desired signal levels are set at $P_{int} = -35$ dBm and $P_{sig} = -70$ dBm, respectively; therefore, the maximum IIP_3 is -20.3 dBm, using a 10 dB margin. The constraints and objectives for the receiver synthesis are shown in Table 6.6 and the optimization results are illustrated in Fig. 6.11.

By comparing Figs. 6.10 and 6.11, it is possible to observe that the Bluetooth POF achieves designs with lower area and power consumption. However, since the Bluetooth standard is more relaxed in terms of specifications when compared to Wi-Fi, it is understandable that this occurs. Furthermore, the Wi-Fi POFs for examples 1 and 2 are very similar because the Wi-Fi standard is very strict in terms of NF. Therefore, in order to comply with the NF constraint, high gain LNAs must be used

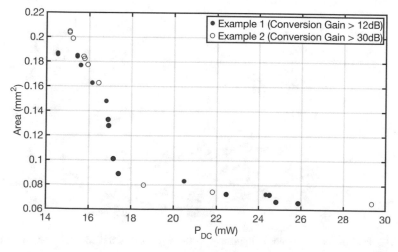

Fig. 6.11 POF of the front-ends compliant with the Wi-Fi standard, for both examples in Table 6.6

in order to reduce the mixer noise (see Eq. (6.4)). Therefore, even though in example 1, the constraint is 12 dB, most of the front-end designs get, in fact, more than 30dB of gain.

After obtaining the POFs with two objectives, is possible to perform another optimization considering more objectives. As a way to consider all the important performances of the front-end receiver, another optimization was considered where a figure of merit (FOM) for the front-end is considered as one of the objectives. The FOM can be defined as

$$FOM = 10 \cdot \log_{10} \frac{10^{CG/10} \cdot 10^{(IIP_3-10)/20}}{10^{NF/10} \cdot P_{DC} \cdot V_{DD}} \tag{6.11}$$

which relates all the performances of the front-end. This FOM can be included in the optimization process as one objective subject to maximization. Thus, two optimizations were performed (one for each standard) where the FOM was included as one objective. The optimizations were performed with 200 individuals and 80 generations and the constraints are the same as those specified for example 2 in Table 6.6. The optimization results can be seen in Figs. 6.12 and 6.13.

Figures 6.10, 6.11, 6.12, and 6.13 show hundreds of fully designed RF front-end receivers with different performance trade-offs, which were automatically designed with the proposed methodology. From the optimization results shown in Figs. 6.12 and 6.13, four individual front-end designs were selected and their performances as well as the performances for each low-level circuit are illustrated in Table 6.7, where the performances in bold are for the entire front-end.

Each of the fully designed RF front-end receivers shown from Figs. 6.10, 6.11, 6.12, and 6.13 can be simulated in any graphical circuit simulator, e.g., Cadence Virtuoso and its performances can be inspected in detail. Therefore, in order to show the performances of a given front-end, a design compliant with the Bluetooth

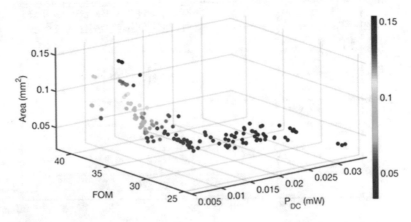

Fig. 6.12 POF of the front-ends compliant with the Bluetooth standard while maximizing the receiver FOM

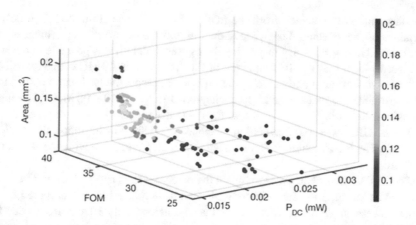

Fig. 6.13 POF of the front-ends compliant with the Wi-Fi standard while maximizing the receiver FOM

Table 6.7 Performances of the individual front-ends (and composing circuits) for the Bluetooth and Wi-Fi standard (for example 2)

Performances	Bluetooth (IEEE 802.15.1)		Wi-Fi (IEEE 802.11b)	
	Design 1	Design 2	Design 1	Design 2
S_{11} (dB)	−17.375	−15.628	−12.637	−12.595
S_{21} (LNA) (dB)	11.785	10.970	15.501	13.956
CG (MIX) (dB)	21.669	21.669	17.472	17.414
CG (dB)	**30.062**	**30.622**	**34.567**	**30.523**
NF (LNA) (dB)	3.750	3.358	2.523	2.329
NF (MIX) (dB)	14.493	14.493	7.527	7.585
NF (dB)	**7.263**	**7.410**	**5.203**	**5.134**
IIP_3 (LNA) (dBm)	−2.8845	−0.164	−3.371	−0.077
IIP_3 (MIX) (dBm)	5.892	5.892	−2.513	−4.880
IIP_3 **(dBm)**	**−9.5173**	**−5.193**	**−16.675**	**−17.790**
PN @ 1MHz (VCO) (dBc/Hz)	−121.940	−119.710	−119.81	−117.92
P_{OUT} (VCO) (dBm)	−8.866	−3.353	−6.830	−1.572
P_{DC} (LNA) (mW)	3.897	19.382	11.915	15.085
P_{DC} (MIX) (mW)	0.749	0.749	1.272	1.523
P_{DC} (VCO) (mW)	2.547	13.115	1.988	12.167
P_{DC} **(mW)**	**7.158**	**33.235**	**15.059**	**28.735**
FOM	**30.513**	**26.420**	**30.269**	**22.930**
Area (LNA) (mm²)	0.0677	0.0215	0.1289	0.0627
Area (VCO) (mm²)	0.0428	0.0102	0.0413	0.0102
Area (MIX) (mm²)	0.8888×10−3	0.8888×10−3	0.0020	0.0019
Area (mm²)	**0.1114**	**0.0327**	**0.1722**	**0.0748**

The bold values are the performances of the entire system while the non-bold are the performances of the sub-circuits.

standard randomly selected from the POF in Fig. 6.12 was simulated. The design variables of the selected design are shown in Table 6.8, and its performances can be observed in Fig. 6.14. By inspecting the figures in detail, it is possible to observe in Figs. 6.14a, b that the IF frequency of the front-end is 40 MHz. In Figs. 6.14c, d, the conversion gain and noise figure are shown, respectively. It can be seen that the constraints are met for an IF frequency-band range of 10–100 MHz. The input matching, shown in Fig. 6.14e, is less than -12 dB for entire ISM frequency band of 2.4–2.5 GHz, where the Bluetooth standard operates. Furthermore, the IIP_3 is shown in Fig. 6.14f, where it can be seen that it also complies with the Bluetooth standard.

The total CPU time in order to obtain the receiver POFs can be seen in Fig. 6.15. The simulation time for the inductor POFs can be neglected when compared to the entire process, since the surrogate model is so efficient, only 10 min are needed for the POF generation. Furthermore, the simulation time for the low-level blocks is around 6 h for the LNA, 9 h for the VCO, and 22 h for the mixer (this will vary depending on the number of generations and individuals that the user defines). One of the advantages of this methodology is that the optimization of these three individual blocks (LNA, VCO and mixer) is completely independent one from another, and therefore, can be run in parallel, minimizing therefore the total CPU time. Also, with the hierarchical reusability of the methodology, they only need to be performed once. After the low-level POFs are obtained, their POFs are mapped into matrices and used in the high-level optimizations.

6.4   Alternative Decomposition Strategies in Receiver Front-End Synthesis

In Sect. 4.2, comparisons were made between hierarchical optimization-based methodologies and methodologies without hierarchical decomposition. The experiments were conducted between circuit and device level. The main difference between both strategies was how to design the inductors: in an online fashion (during the optimization) or using an inductor POF, which means that the inductors were designed in an offline fashion.

In this section, these concepts will be applied and compared for the design of the entire RF front-end receiver. While in Sect. 4.2 the comparisons were made between the circuit and device level, here, it is also interesting to perform these comparisons at the system level. Therefore, several different optimizations were performed. The optimizations were performed for the previously considered Bluetooth and Wi-Fi standards (considering the specifications of example 1 in Table 6.6). The system-level hierarchical optimization, performed in the previous section, is illustrated in Fig. 6.16 for convenience, and it will further be denoted as BU for the sake of simplicity.

Table 6.8 Design variables for one individual of the front-end POF (individual performances illustrated in Fig. 6.14)

Circuit	Design variables										
LNA	W_1 (µm)	W_2 (µm)	$l_{1,2}$ (µm)	V_b (V)	C_1 (pF)	C_2 (pF)	C_3 (pF)	L_S	L_G	L_B	L_D
	510	300	0.35	0.643	0.6	3.96	1.12	N=2	N=7	N=2	N=6
								D_{in}=10 µm	D_{in}=42 µm	D_{in}=205 µm	D_{in}=29 µm
								w=5.15 µm	w=5 µm	w=12.4 µm	w=5.05 µm
VCO	W_N (µm)	W_P (µm)	$L_{N,P}$ (µm)	I_bias (µA)	C_{LO} (pF)	C_{var} (µm)	Row_{Cvar}	Col_{Cvar}	W_{CM-D} (µm)	W_{CM} (µm)	L
	20	140	0.35	420	0.4	Width=105.6	6	7	60	130	N=4
						Length=0.65					D_{in}=128 µm
											w=7.85 µm
MIX	W_{LO} (µm)	W_{RF} (µm)	W_{CM-2} (µm)	$L_{LO,RF,CM-2}$ (µm)	I_{bias} (µA)	R_{MIX} (KΩ)	R_{CHOKE} (KΩ)	C_{MIX} (pF)			
	20	50	130	0.35	500	4.71	18.66	0.3			

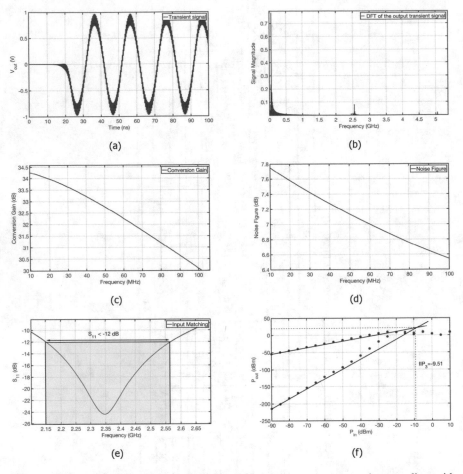

Fig. 6.14 Front-end performance for the design with lowest power consumption compliant with the Bluetooth standard from the obtained POF in Fig. 6.10. (**a**) Transient simulation. (**b**) DFT simulation. (**c**) Conversion gain. (**d**) Noise figure. (**e**) Input matching. (**f**) Third-order interception point

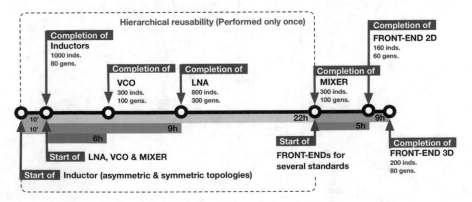

Fig. 6.15 Timeline of the entire proposed design methodology for the design of RF systems

OPTIMIZATION AT SYSTEM LEVEL
CIRCUITS AND PASSIVES PASSED AS POFS

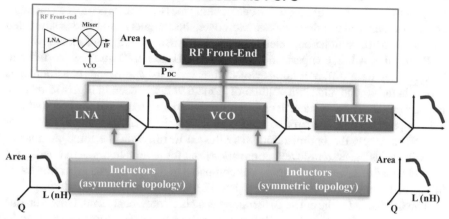

Fig. 6.16 Illustrating the bottom-up (BU) hierarchical optimization

OPTIMIZATION AT DEVICE LEVEL
CIRCUITS AND PASSIVES OPTIMIZED TOGETHER

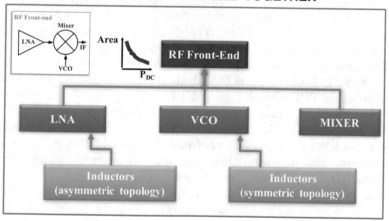

Fig. 6.17 Illustrating the device-level (IND) optimization

6.4.1   Device-Level Optimization

In this sub-section, a device-level optimization is going to be performed. This optimization is considered at device level, because all circuits and passives are sized during the optimization. Further on, this optimization will be denoted as IND for the sake of simplicity and its strategy can be observed in Fig. 6.17. The IND strategy is the most straightforward optimization strategy because no hierarchical decomposition is performed. Thus, the design search space includes all the design variables shown in Table 6.1 for the inductors and in Tables 6.3, 6.4, and 6.5 for the circuits. During the optimization, SIDe-O is used in order to model both the asymmetric and symmetric inductors, calculating the S-parameters for each inductor considered during the sizing process. Moreover, the complete front-end is simulated with an electrical simulation, calculating its performances with high accuracy.

Performing a fair comparison between the BU and IND strategy is difficult, because they have different number of design variables and design constraints, as shown in the second and sixth column of Table 6.9. It was seen in Fig. 6.15 that the BU strategy needed around 42 h of CPU time in order to reach the 2D POF (adding the computation time needed to optimize all the passives, circuits, and system). For the IND strategy, the optimization was allowed to run with the same number of individuals (as the system-level optimization) and for several hours. Afterwards, the POF obtained at several CPU times was compared to the BU approach, as illustrated in Fig. 6.18, for the Bluetooth standard and Fig. 6.19 for the Wi-Fi standard.

The CPU time for the entire BU strategy is 42 h, as previously said. However, due to the low-level hierarchical reusability, the optimization of the passives and circuits only has to be performed once. Therefore, the actual system synthesis is only 5 h CPU time (as shown in the legend of Figs. 6.18 and 6.19). In both Figs. 6.18 and 6.19, the BU POF can be seen in red dots. In green dots it is possible to observe the IND POF at generation 200, where the consumed CPU time was around 28 h. It is possible to observe that at this generation there is a cloud of points, and no actual POF (for both Bluetooth and Wi-Fi). Therefore, the optimization was allowed to evolve up to 400 generations (black dots in Figs. 6.18 and 6.19), extending to a total CPU time of 55 h. Despite the high CPU time, it is possible to conclude that when using the IND strategy, the POFs obtained for both standards are completely dominated by the POFs obtained using the BU strategy, therefore, endorsing the usage of multilevel BU strategies for the design of RF circuits.

From the optimization times shown in Figs. 6.18 and 6.19, it may be perceived that there is a lack of proportionality between the BU and IND CPU optimization times. The exact same schematic is being simulated, where the BU strategy performs 9600 front-end electrical simulations (160 individuals times 60 generations), and the IND strategy performs 64,000 simulations (160 individuals times 400 generations). In the IND strategy we are performing (roughly) six times the number of simulations but the optimization time is more than six times that of the BU strategy. Therefore, in order to understand this, some drawbacks of the IND strategy must be pointed out:

- In the IND strategy, the inductors are designed during the optimization. Since SIDe-O is being used, the S-parameter files have to be created at each generation (as explained in Chap. 4), whereas in the BU strategy these files are created a priori. The time needed to evaluate the inductors with SIDe-O and create their S-parameter files degrades the efficiency of the process.
- The IND strategy has to perform more simulations at each generation in order to comply with all front-end constraints. In order to evaluate the front-end input matching, an S-parameter analysis must be performed, and, in order to consider

Table 6.9 Comparison between the constraints of the different optimization strategies. Illustrating the total number of design variables and constraints

Performances	BU Bluetooth	BU Wi-Fi	CIR Bluetooth	CIR Wi-Fi	IND Bluetooth	IND Wi-Fi
S_{11}	x^a	x	<12 dB	<12 dB	<12 dB	<12 dB
CG	>12 dB	>12 dB	>12 dB	>12 dB	>12 dB	>12 dB
NF	<8.79 dB	<5.64 dB	<8.79 dB	<5.64 dB	<8.79 dB	<5.64 dB
IIP_3	>−10.35 dBm	>−20.3 dBm	>−10.35 dBm	>−20.3 dBm	>−10.35 dBm	>−20.3 dBm
f_{osc} (VCO)	x	x	>2.45 GHz	>2.45 GHz	>2.45 GHz	>2.45 GHz
f_{osc} (VCO)	x	x	<2.55 GHz	<2.55 GHz	<2.55 GHz	<2.55 GHz
Port-to-port isolation	x	x	>30 dB	>30 dB	>30 dB	>30 dB
Inductor constraints	x	x	x	x	Yes^b	Yes
P_{DC}	Minimize	Minimize	Minimize	Minimize	Minimize	Minimize
P_{DC}	<40 mW	<40 mW	<40 mW	<40 mW	<40 mW	<40 mW
Area	Minimize	Minimize	Minimize	Minimize	Minimize	Minimize
Optimization settings						
Number of design variables	9^c	9	33	33^d	38^e	38
Number of constraints	6	6	15	15	40	40

[a] The constraints marked with an x were already imposed at circuit/passive level and do not need to be imposed at circuit/system level

[b] The inductor constraints are the ones defined in Chap. 3, in order to ensure that the inductors are in the plain-BW zone (4 constraints per inductor), plus the inductor area constraint (D_{out} <400 μm)

[c] The design variables of the BU approach are the indexes of the matrix mapping for each circuit

[d] All design variables of the LNA, VCO, and mixer are considered. The inductors are passed as indexes from the matrix mapping

[e] All design variables of the LNA, VCO, and mixer are considered, plus the geometrical parameters of inductors

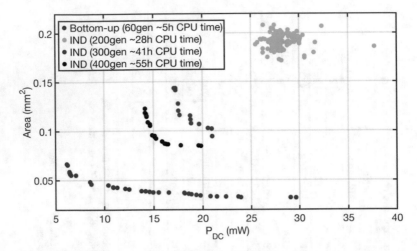

Fig. 6.18 Comparison between the POFs obtained with the BU and IND strategies for the Bluetooth standard. (Reprinted with permission [1])

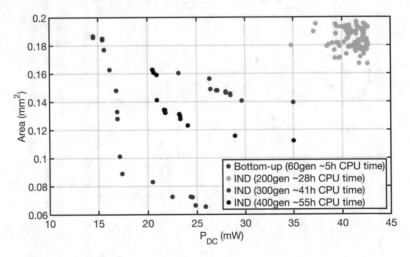

Fig. 6.19 Comparison between the POFs obtained with the BU and IND strategies for the Wi-Fi standard. (Reprinted with permission [1])

the port-to-port isolation, a PXF analysis must also be performed. In the BU strategy, these analysis are performed during the simulation of the LNA (the input matching) and the mixer (the port-to-port isolation), relieving, therefore, the front-end optimizations from these simulations.

- Furthermore, one of the analysis of the front-end is the PSS. The time needed for this simulation highly depends on the individual design being simulated, because the analysis needs to converge to a steady-state: different individuals may lead to different simulation times. While in the BU strategy, every single

LNA, VCO, and mixer designs are fully designed and work properly, in the IND strategy, during the initial stages of the optimization, there may be plenty of, e.g., VCOs that take longer times to converge. Therefore, in the initial stages of the optimization, these PSS simulations may last longer, hampering the optimization efficiency.

From the POFs shown in Figs. 6.18 and 6.19 and the previous given drawbacks of the IND strategy, it may be concluded that the BU strategy is more efficient and achieves superior results when compared to the IND strategy.

6.4.2   Circuit-Level Hierarchical Optimization

The optimization at device level (IND strategy in Sect. 6.4.1) has the drawback of designing the inductors in an online fashion, highly increasing the number of design variables and constraints imposed in the IND strategy. Therefore, it may be possible that the optimization algorithm struggles to converge to optimal solutions.

In order to study the effect of such online inductor design in the entire front-end optimization, a circuit-level hierarchical optimization was performed, where the circuits composing the front-end (LNA, VCO, and mixer) are sized during the same optimization. However, the inductors are designed offline and passed as a POF. This optimization will be denoted as CIR for the sake of simplicity and the strategy can be observed in Fig. 6.20. Furthermore, the number of design variables and design constraints are shown in the fourth and fifth column of Table 6.9.

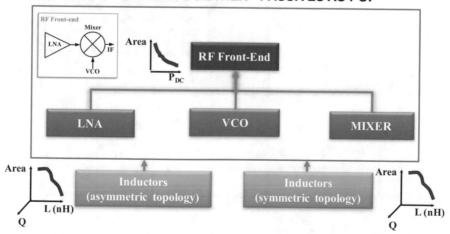

Fig. 6.20 Illustrating the circuit-level (CIR) optimization

Again, since a fair comparison between BU and CIR strategies is very difficult, the optimization was run with the same number of individuals but for a long CPU time in order to inspect the convergence of the optimization. The results of the optimization can be seen in Fig. 6.21 for the Bluetooth standard and in Fig. 6.22 for the Wi-Fi standard.

The front-end POF obtained with the BU strategy is illustrated by red dots in both Figs. 6.21 and 6.22. In green dots it is possible to observe the CIR POF at generation 200, where the consumed CPU time was around 25 h. It can be seen that the obtained

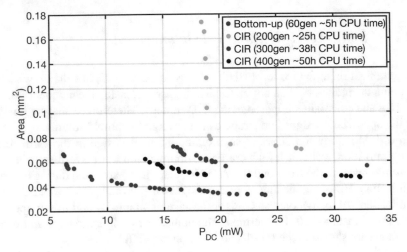

Fig. 6.21 Comparison between the POFs obtained with the BU and CIR strategies for the Bluetooth standard. (Reprinted with permission [1])

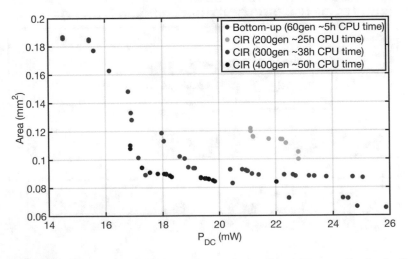

Fig. 6.22 Comparison between the POFs obtained with the BU and CIR strategies for the Wi-Fi standard. (Reprinted with permission [1])

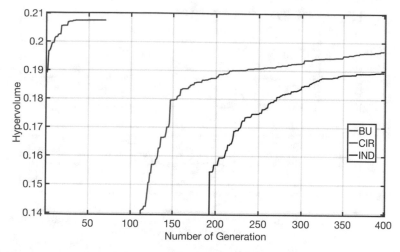

Fig. 6.23 Hypervolume metric over each generation for the Bluetooth standard

POF is still far away from the one obtained with the BU strategy. The optimization was run up to 50 h of CPU time, where the obtained POF is shown in black dots in Figs. 6.21 and 6.22. For the Bluetooth standard, the results are similar to the IND strategy, where despite the high CPU time of the CIR approach (with 50 h of CPU time), the obtained POF is completely dominated by the POF obtained with the BU strategy, therefore, endorsing the usage of multilevel BU strategies for the design of RF circuits. For the Wi-Fi standard, shown in Fig. 6.22, the CIR optimization with 400 generations slightly overlaps the BU POF; however, the BU POF is much wider and achieves lower power consumptions and lower areas.

It is also interesting to observe the hypervolume metric over each generation in order to understand if more generations could help the convergence of all strategies. In Figs. 6.23 and 6.24, the hypervolume is shown for each generation for both standards and for all presented strategies. The hypervolume metric depends on the reference point; hence, the same reference point was used for all techniques and during all generations in order to fairly compare the Pareto fronts generated with all the different strategies.

Regarding the hypervolume curves, the hypervolume is zero when no feasible solution is available on the POF. Therefore, it is possible to observe that the BU method has feasible solutions since early generations, whereas the CIR and the IND method need many more generations in order to meet the optimization constraints. In Fig. 6.23, the hypervolume for the Bluetooth experiments is shown. It is possible to observe that the hypervolume curves are quite stable at the final generations and therefore more generations would probably not significantly improve the POF quality. In Fig. 6.24, the hypervolume curves for the Wi-Fi experiments are shown. In this figure, the hypervolume is still growing for the IND method, and, therefore, more generations would probably increase the quality of the POF. However,

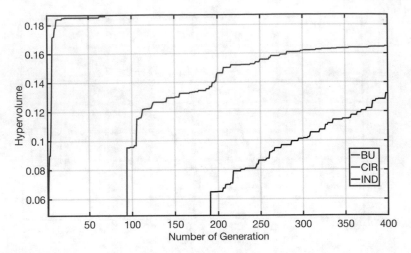

Fig. 6.24 Hypervolume metric over each generation for the Wi-Fi standard

increasing the number of generations would increase the CPU time, leading to unpractical simulation times (more than 55 h of CPU time for each experiment).

From these experiments, it is possible to conclude that bottom-up design methodologies, decomposing the problem in a hierarchical fashion, prove to be a more efficient method to design RF systems. Furthermore, with these BU strategies, and, due to the hierarchical reusability, obtaining the POF for another communication standard would be very fast (only 5 h), whereas for the IND and CIR methods, the same results are impossible to obtain with such efficiency.

6.5   Summary

In this chapter a multilevel BU circuit design methodology has been described and applied to the design of an RF system composed of an LNA, a VCO, and a mixer. By using such multilevel BU strategy, different circuits can be merged to design an RF system. Furthermore, each level of the hierarchy is simulated with the utmost accuracy possible: EM accuracy at device level and electrical simulations at circuit and system level. Also, the methodology used promotes the hierarchical reusability of low-level POFs. Moreover, the methodology proved to be highly efficient and presented superior results when compared to other alternative partitioning and synthesis strategies.

References

1. F. Passos, E. Roca, J. Sieiro, R. Fiorelli, R. Castro-López, J.M. López-Villegas, F.V. Fernández, A multilevel bottom-up optimization methodology for the automated synthesis of RF systems. IEEE Trans. Comput.-Aid. Des. Integr. Circuits Syst. **39**, 560–571 (2020). https://doi.org/10.1109/TCAD.2018.2890528
2. H. Friis, Noise figures of radio receivers. Proc. IRE **32**(7), 419–422 (1944)
3. R. Li, *RF Circuit Design* (Wiley, Hoboken, 2002)
4. T. Eeckelaert, R. Schoofs, G. Gielen, M. Steyaert, W. Sansen, An efficient methodology for hierarchical synthesis of mixed-signal systems with fully integrated building block topology selection, in *Design, Automation Test in Europe Conference Exhibition* (2007), pp. 1–6
5. A.A.E. Emira, Bluetooth/WLAN receiver design methodology and IC implementations. PhD Thesis, EECS Department, Texas A & M University (2003)
6. P. van Zeijl, J.W.T. Eikenbroek, P.P. Vervoort, S. Setty, J. Tangenherg, G. Shipton, E. Kooistra, I.C. Keekstra, D. Belot, K. Visser, E. Bosma, S.C. Blaakmeer, A Bluetooth radio in 0.18-μm cmos. IEEE J. Solid-State Circuits **37**, 1679–1687 (2002)
7. M. Chen, K.H. Wang, D. Zhao, L. Dai, Z. Soe, P. Rogers, A CMOS Bluetooth radio transceiver using a sliding-IF architecture, in *IEEE Custom Integrated Circuits Conference* (2003), pp. 455–458

Chapter 7
Conclusions

The design of RF circuits is highly complex, with difficult to manage and convoluted trade-offs. The usage of systematic design methodologies can help to design these RF circuits more efficiently, hence reducing the time-to-market. The main objective of this book was to illustrate new systematic design methodologies capable of improving the state-of-the-art, cutting short the distance between the RF and digital computer-aided design tools. By doing so, it is possible to reduce the time-to-market and shorten the existing productivity design gap in RF circuit design. The methodologies shown in this book tackle several bottlenecks of the RF design.

First, a state-of-the-art surrogate modeling strategy for integrated inductors was developed for the accurate and efficient modeling of integrated inductors. The presented model shows less than 1% error when compared to EM simulations while reducing the simulation time by three orders of magnitude. Several models were created for different inductor topologies. The accuracy and efficiency of the surrogate model developed enable its usage within iterative optimization loops. In this book, several inductor synthesis strategies were compared. From this comparison, it was possible to conclude that building an accurate global surrogate model shows very significant advantages when compared to state-of-the-art surrogate-assisted optimization strategies (e.g., ONSO and ONSOEI). Furthermore, a tool for the design and optimization of integrated inductors, SIDe-O, was developed. This tool offers an intuitive graphical user interface (GUI), which can be used by an RF designer in order to model, simulate, and design integrated inductors for different topologies, operating frequencies, and technological processes. Furthermore, SIDe-O also allows the creation of S-parameter files that accurately describe the behavior of inductors for a given frequency range, which can later be used in electrical simulations for circuit design in commercial environments. The surrogate models developed, and integrated in the tool, provide a solution to the problem of accurately and efficiently modeling inductors, as well as their optimization, alleviating the bottleneck that these devices represent in the RF circuit design process.

© Springer Nature Switzerland AG 2020
F. Passos et al., *Automated Hierarchical Synthesis of Radio-Frequency Integrated Circuits and Systems*, https://doi.org/10.1007/978-3-030-47247-4_7

Regarding the RF circuit design methodologies, in this book, a wide study was first performed in order to compare two different optimization-based RF design methodologies: one based on hierarchical decomposition and bottom-up synthesis and another where no hierarchical decomposition was performed and synthesis was performed at the circuit level. The study was performed between the lowest possible level: device level and the immediately superior level: circuit level. It was demonstrated that bottom-up hierarchical design methodologies are far more efficient and are able to achieve superior results. Therefore, in this book, hierarchical bottom-up design methodologies were supported and applied to the design of different RF circuits, such as LNAs and VCOs. The methodology uses the SIDe-O tool in order to generate Pareto fronts of inductors, which can later be used in circuit simulations. Gilbert cell mixers were also designed using systematic design methodologies; however, since the Gilbert cell mixer does not have inductors, its design was considered using an optimization-based strategy without hierarchical decomposition. Furthermore, in all circuit optimizations, several simulation strategies were used in order to reduce the circuit simulation time. By using such strategies some of the most expensive RF performances (e.g., third-order intercept point) can be efficiently calculated and considered during the automated design of RF circuits.

Another important issue in RF circuit design are the layout parasitics that affect the circuit performances during the physical synthesis. Since these parasitics are so destructive, a layout-aware methodology was developed specifically for the design of RF circuits. The methodology uses multi-objective optimization algorithms and a bottom-up design methodology in order to design the circuits. In this book, this methodology was applied to the design of an LNA, a VCO, and a mixer. An automatic layout generation is performed during the optimization for each sizing solution using a state-of-the-art module generator, template-based placer and router, which were specifically developed for RF circuits. The proposed approach exploits the full capabilities of most established computer-aided design tools for RF design available nowadays, i.e., the RF circuit simulator as performance evaluator and commercial layout parasitic extractor to determine the complete circuit layout parasitics. Furthermore, in order to accurately address the parasitics of each device, the inductor parasitics are considered using the SIDe-O tool. The methodology developed also allows the user to consider the corner performances not only during the sizing optimization, but also during the layout-aware optimization, increasing, therefore, the design robustness.

The circuit complexity which can be tackled by the methodology is a key issue for its validity. In this book, a multilevel bottom-up circuit design methodology was described and applied to design an RF system composed of three blocks (an RF front-end composed of an LNA, a VCO, and mixer). By using such multilevel bottom-up strategy, different circuits can be connected in order to build an RF system. Furthermore, each level of the hierarchy is simulated with the upmost accuracy possible: EM accuracy at device level and electrical simulations at circuit/system level. The methodology developed in this book encourages the hierarchical low-level POF reuse, which is typical in bottom-up methodologies.

Moreover, the methodology proved to be highly efficient for the design of RF front-ends for different communication standards. When compared to alternative synthesis strategies for RF systems, the presented methodology shows superior results.

In summary, this book presented a multilevel approach to design RF circuits and systems, where the system is designed in a bottom-up fashion, starting from the device-level stage. Furthermore, at each stage of the design hierarchy, several aspects are taken into account in order to increase the design robustness and increase the accuracy and efficiency of the simulations.